Horses in Harness

By Charles Philip Fox

DEDICATION
To my father, Dr. G.W. Fox, Sr.
and
My grandfather, Dr. Philip Fox
Physicians and surgeons of Milwaukee and Madison, Wisconsin, who lived in
the era of horses in harness, and from whom I inherited my love of and great interest in horses.

Editor: Jean Van Dyke
Art Director: Peggy Bjorkman
Art Associate: Cindy Domin
Production: Loretta Caughey, Sally Manich

International Standard Book Number: 0-89821-080-1
Library of Congress Catalog Card Number: 87-62658
©1987, Reiman Associates, Inc. 5400 S. 60th St., Greendale WI 53129.
All rights reserved. Printed in U.S.A.

CONTENTS

INTRODUCTION

IN THE 1920s, when we were living in Madison, Wisconsin, we would visit my grandfather. Naturally, our first question would be, "How are you feeling, Grandpa?" His answer was a vivid description of his spirits. "Oh, head high, tail over the dashboard," would come his reply.

Once in a while, he would cross us up by saying, "I am not feeling my oats today." This was taken as a signal for a short visit that particular day.

I don't remember much more of the visits than those comments. After all, a 10- or 12-year-old kid was supposed to be seen, not heard. But my already well-established interest in horses made my grandfather's words memorable.

My inherent interest in horses, in my case, draft horses, expressed itself in other ways as well. I began to build scrapbooks about horses, to photograph with my Brownie 2A Eastman box camera the horses still on the city streets and to spend hours each year in the horse stables at the Dane County Fair.

The day of days each summer was when a circus came to town. If it was the Sparks Circus or Hagenbeck-Wallace Circus they might have 100 to 150 draft horses. But when Ringling Bros. and Barnum & Bailey Combined Shows arrived with 250 dapple-grey Percherons, it was paradise! The performance in the Big Top was secondary to me. The real show, the important show, was watching the matched 6- and 8-horse teams work.

The draft horse was fast disappearing from the city streets by the 1920s. By the following decade, nearly all the horses left were pulling peddlers' wagons and milk wagons.

I decided to compile this pictorial history long after the horses were gone. I wanted to put together one book to show the present generations how things were done when using horses was the only way to accomplish most tasks. This book shows, primarily with photographs, how horses delivered the merchandise, took people where they needed to go, hauled just about anything that needed to be moved and did the jobs that needed to be done.

A great deal of fascinating information is drawn from the *Breeder's Gazette,* "The Weekly Journal of Livestock Husbandry." Back in 1940 a dear friend, Hugh Van Pelt, breeder of Jersey cattle in Waterloo, Iowa, gave me a set of bound volumes of the *Breeder's Gazette.* Those issues covered the period from 1892 to 1921—the last 30 years when the horse reigned supreme. Photos and information from the Gazette help illustrate that horses did indeed get things done—and with dispatch.

When all is said and done, this book is a tribute to the horses that worked in harness and helped pull America to the forefront. It is not intended to be definitive. There are complete books available on many of the subjects, such as vehicles, fire horses, circus and farming.

There has been a definite surge in interest in draft horses over the last 15 years. A new group of people is carrying on old traditions at county and state fairs, with pulling contests, plowing contests, the multiple horse hitches, parades and more.

I hope this book will be an inspiration for the horsemen and women of today when they see the astounding, remarkable and at times unbelievable chores accomplished by their forefathers and their draft horses.

—*Charles Philip Fox*

Prehistoric man, dwelling in earth's huge caverns, preserved a record of the most notable achievement of his age, of the noblest conquest ever made by man over the brute creation. Upon the walls of his subterranean home, carved in the imperishable rock, amid rude sketches of mastodons, of cave bears, of reindeers and other objects of his dread or of the chase, again and yet again man draws the picture of a bridled horse.

Before kingdoms were conceived, before social order was known, before tribal law was recognized, horse and man proclaimed the coming civilization. The domestic tools of the earliest agriculturists and the weapons of the first warriors are ornamented with the head of the haltered horse.

Man's first flights of imagination were inspired by the horse. The centaur was man's ideal warrior, a horse in pride, strength, speed, agility and courage, loving danger and seeking it with ardor, guided by the brain and equipped with the arms and weapons of man. The wooden horse of Troy, the flying horse of the Arabian Nights, Pegasus, and the Four Horsemen of the Apocalypse are feeble acknowledgments of the intellectual indebtedness of man to the horse.

Man has ever paid homage to the instinct of his horse as a sense more highly developed than the reasoning powers of man.

Inventive genius was aroused by the horse.

The horse was inspiration to troubadour and minnesinger and has been, is and will be for all time to painters, to poets and to sculptors.

Man freely confesses his obligation to his horse, his love and affection for his horse and that every noble attribute man possesses is found in its perfection in his horse.

Man in all ages has proudly likened himself unto his horse, has gladly lived with him under the same tent as companion of his journeys, cheerfully toiled with him as a servant in the fields, hunted with him as his comrade, raced with him as the fleetest beast afoot, fought with him as his truest and fiercest ally, died and been buried with him, begging with his last breath that his horse be sculptored on his tomb.

Together they have endured the privations and hardships of toil; together they have shared the dangers and glory of war.

Together they shall enjoy the fruits of their labors and together divide the honors of eternal peace.

Tribute to the horse, written in 1924 by George E. Wentworth, then superintendent of the horse market department of the Union Stock Yards, Chicago.

ACKNOWLEDGMENTS

**Extraordinary assistance
has been received from these people which
I gratefully acknowledge:**

Wesley W. Jung, Sheboygan, Wisconsin
Gene Baxter, Troy, New York
Philip Weber, Cleveland, Ohio
(co-author, *Heroes in Harness*)
Mrs. Barbara McKellar, Larkspur, Colorado
Edward F. Tracy, Middletown, Rhode Island
Tom Ryder, Salem, New Jersey
(author, *On the Box Seat*)
Bud Dillner, Orlando, Florida
H.S. Walker, Marblehead, Massachusetts
George F. Getz, Jr., Scottsdale, Arizona
Ron Ryder, Boonville, New York
Richard Mueller, Milwaukee, Wisconsin
Norman Coughlin, Chicago, Illinois
Jerry Booker, San Mateo, California
Phil Fox, Oregon, Wisconsin
Arnold Hexom, Waverly, Iowa
Bud Yarke, Huntley, Illinois
Dr. H.J. McGinnis, Waupaca, Wisconsin

CHAPTER I

BEGINNING OF THE END

THROUGHOUT the 19th century, inventors, tinkers, and mechanics cobbled together all types and styles of automated carriages they thought would push the horse into oblivion.

These contraptions were propelled with oil-fired steam engines, compressed air, and vapor (gas) engines. Weird in design, and impractical, they were ahead of their time.

As the 19th century drew to a close, things seemed to jell as designs improved. Horseless vehicles met in stiff competition in France in 1894, with 102 entries in the 68-mile run from Paris to Rouen. Included in the race were electric vehicles, as well as those that ran on steam, petroleum, or compressed air (*Carriage Monthly*, September 1894).

Two years later the Duryea automobile became a circus attraction when the Barnum & Bailey Greatest Show On Earth carried one around the country on their summer tour.

In each town the Circus had the auto driven down the main thoroughfare as part of their free street parade. During the performance under the big top it putt-putted around the hippodrome track for all towners to see "the coming mode of travel".

That season Barnum & Bailey scheduled 139 towns from New York City through New Jersey, New England, Pennsylvania, Ohio,

AT THE FAIR GROUNDS. Before the the turn of the century, horses were the center of interest at county and state fairs. They were featured exhibits and furnished key entertainment, as seen in this 1894 Currier & Ives print.

BEGINNING OF THE END

Indiana, and closing in Chicago. People in towns such as Pottsville, Pennsylvania; Poughkeepsie, New York; Ansonia, Connecticut; Pawtucket, Rhode Island; Holyoke, Massachusetts; Brattleboro, Vermont; Biddeford, Maine; Troy and Buffalo, New York; Galion, Ohio and dozens of other communities probably saw their first automobile at the circus.

Early in the 20th century, crazy rules and regulations popped up in towns in an effort to get control over the autos. One such law required the driver of a car to put a canvas painted with country scenery over his auto if a horsedrawn vehicle refused to go by. At night, the motorist was required to send up a rocket every mile and wait for ten minutes for the road ahead to be cleared of horse traffic before proceeding.

(The drivers and riders of horses had their share of regulations, too. New York State had a statute on the books that said it was unlawful to drive, ride, or lead a horse over any bridge faster than a walk. The fine was $15.00 and the arrested person had to pay for any damage to the bridge. Some towns, in an effort to cut down on aimless galloping around and general letting off of steam, passed such ordinances as ``it shall be unlawful to drive or ride a horse or mule at a faster gait than six miles per hour within the city limits''.)

Like the tide rolling in, however, the horseless carriage was gradually taking over. One authority pointed out it was really the development of the bicycle that made people realize they could get around town in faster and easier ways than using horse-drawn street cars, or having the expense and trouble of owning a horse and buggy.

In 1907 the Kiblinger automobile was advertised as, ``A high grade machine at a fair price.'' It had an air-cooled 2-cylinder engine capable of moving the vehicle at 30 MPH,

``THE STATUE OF INDUSTRY''. Two of these statues flanked the main entrance to the Palace of Mechanic Arts at the 1893 Columbian Exposition in Chicago. Magnificent monuments to the draft horse, harness and all, were rare even in the era when they did all the work.

Credit: Author's Collection

THE COMING MODE OF TRAVEL. It took a traveling circus to introduce the auto to the people of the hinterland, as seen in this 1896 poster.

HORSE WAS KING. The 1905 International Livestock Exposition used this artwork to advertise the coming show.

Credit: Author's Collection

solid rubber tires and "top if desired"—all for $375.00. In their ad, the Holsman automobile claimed they were the "pioneer of high-wheeled autos for summer and winter." This same year the Pope-Toledo, Columbia and the Haynes-Apperson pushed their products. A year later ads were seen for the Rambler, Cartercar, Schact Runabout and Black Motor Buggy.

Some tried to stem the tide by denying the takeover by the auto.

"The horseless age has existed in the daily papers for many years. It has existed nowhere else and never will be found anywhere else.

"The horse is in no more peril of extinction by the automobile than he is of being driven into oblivion by the airship.

"We are no nearer to the horseless age than the manless age." (*Breeders Gazette*, Nov. 24, 1909)

Horse dealers had conflicting opinions. "F. M. Ware, Manager of the American Horse Exchange, claims to have arrived at the conclusion that the horse has been superseded by the automobile and it will be but a short time before there will not be a public horse market in New York City." (*Breeders Gazette*, Nov. 24, 1909)

Yet, in the same magazine, "The great establishment of Fisk, Doerr and Carroll, horse dealers, reported better business the first six months of 1909 than any similar period in their history".

One thing for certain, the noisy, honking oncoming rush of the auto was aggravating to horsemen. "We do not consider an auto an annoyance, but the annoyance comes from the motorist who is ignorant of his duty to his fellow man. A motorist is supposed to be an engineer who so perfectly understands his machine that he can control it on short notice and avoid accidents. On the other hand, a spirited team is not so easily controlled as horses have a will of their own." (*Breeder's Gazette*, Sept. 28, 1910)

BEGINNING OF THE END

No one fussed when some of the bigger companies began to use three-, four-, and even six-horse wagons, delivering more merchandise on one trip, and replacing two or more of the one- and two-horse outfits.

But, things took a different twist when one truck did the work of two wagons. The *Breeder's Gazette*, in its issue of Oct. 9, 1912, editorialized: "In Chicago a large bottling company put into operation six big trucks, each one doing the work of two teams of horses.

"On refusal of the company to restrict each truck to the work of a single team, the 30 teamsters quit work.

"Such action avails nothing. Wherever autotrucks prove their worth, they will be used—a strike will only serve to encourage a firm to make a complete displacement of teams. A capable teamster could, if necessary, soon fit himself for a position at larger

"TAKE YOUR PEN IN HAND," says the Sears ad in the March 31, 1909, issue of *Breeder's Gazette*. The company was offering a motor buggy "complete and ready to run." The $395.00 price included a top, lamps, fenders, roller-bearing axles, and rubber tires. Another ominous sign for the future of the horse-drawn era was this mail order catalog offer.

SOLID ENTERTAINMENT. Harness races, with pacers or trotters, were held at state and county fairs.

Horse vs. Auto in Chicago.

"The impression most people gather from being in the business section of a city." says Farm Implement News, "these days is that the automobile has become the dominant factor in street traffic. Whether one takes his stand down town, or on the boulevards or in the parks, it is all the same—the automobile seems to dominate the scene. That it is very much in evidence there is no doubt. In Chicago, for instance, traffic experts say automobiles daily in the loop district have increased in number by 600 per cent. during the last five years. This in a way accounts for the popular delusion that the horse is disappearing from the city streets. Upon the contrary, actual investigations of traffic conditions in Chicago show that the number of horse-drawn vehicles in daily service in the down-town district has increased by 30 per cent. during the last two years, and that there are daily 130,000 teams on the central business streets of this city.

"What the automobile truck may do eventually to drive some of these horses off the streets still is problematical; surely up to the present time the use of mechanical trucks has served hardly to check the normal annual increase in the number of horse-drawn vehicles.

"Incidental to this subject, the Rush Street Bridge in Chicago is the most crowded thoroughfare in the world, by actual count 9,000 vehicles of all kinds passing over it every business day, thus taking the reputation of London bridge away, which, with its 7,600 vehicles, is a poor second, although even so the most crowded of any other place in Europe."

AUXILLIARY HORSEPOWER. Sometimes the "good old days" were not so good. This 1903 Rambler did not make it back home to Jefferson, Wisconsin.

THE AUTO TAKES OVER. The October, 1913, issue of the *Blacksmith and Wheelwright* contained this interesting piece.

BEGINNING OF THE END

pay at the steering wheel."

Horsemen, it became gradually evident, were in a losing battle. Seemingly the trend to autos and trucks could not be stopped. But horses were still economically a better bet in some areas of work.

"A survey indicated that 48 leading business establishments in Chicago, including department stores, breweries, and packing houses were using both trucks and draft geldings.

"The general opinion is that trucks are less

The Future of Horse-Drawn Implements.

Tractor makers and dealers have been altogether too sweeping in their statements consigning the work horse to oblivion. It is so refreshing to hear a commonsense statement of fact as between horse and tractor, that we take pleasure in quoting from a monthly house organ issued by the John Deere company to its dealers:

"Have you carefully considered the future of horse-drawn implements? You have built up your business on these implements. Handling them still constitutes a major part of your trade. But now you have a competitor for that trade, and that competitor—your tractor-drawn implement business—is growing pretty fast. You feel the necessity of promoting this new business fully. While you are doing that, however, do not neglect your horse-drawn implement trade. For our friend Old Dobbin has not gone yet. Not by a good deal. He is still there on all four feet, ready to do any farm work. He is a mighty strong favorite with lots of farmers. And he is going to be one of the busiest factors in farm operations for a long, long time yet, maintaining a good demand for good implements. The kerosene-burning tractor of today is all right. Its use will continue to develop. There is a strong future for tractor-drawn implements. Building up a solid foundation for service to farmers who operate tractors is worth all of the energy that you can put into it. But at the same time be sure to give due consideration to the assured place of that faithful, flexible, steady and profitable power-plant—the farm horse."

DON'T FORGET THE HORSE. In 1918 the John Deere Farm Implement Co. felt it necessary to caution their dealers not to neglect the horse-drawn implement business. They forcefully admonished them in their house organ and it so impressed the editors of the *Breeder's Gazette* that they reprinted the article.

Motor Wheel Carriage.

Why not push or propel the ordinary horse-drawn vehicle with a motor wheel. Bicycles are being propelled by them, both single wheels and tandem, and they are so simple of control and so low in cost and in upkeep that the only question is whether one can be built that has the necessary propelling power. But if a one-horse power motor wheel can be purchased at a cost of $60 why not a four or six horse power propelling

Motor Wheel Propulsion.

wheel be sold for about $125, and be as easily attached to the ordinary carriage as it is to the bicycle? The sketch illustrates the idea only. The exact details must be worked out by the practical engineer or mechanic. It shows how a motor wheel of about six horsepower may be attached to an ordinary carriage, it being controlled by a small thumb lever, running from the wheel to the carriage seat as shown by a small flexible cable. This thumb lever regulates the stopping, starting and speed.

BRIGGS AND STRATTON CO. devised this ingenious way to motorize a buggy. This story and sketch appeared in the October, 1915, issue of the *Blacksmith and Wheelwright*.

OVER 39,000 Maxwells were in use when the manufacturer, "feeling his oats", ran this full-page ad. J.D. Maxwell bragged about his "Economy Car being Perfectly Simple—Simply Perfect".

The First Real Cost-Test Ever Made
Automobile *vs.* Horse

"Maxwell" per passenger mile—$1\frac{8}{10}$ cents
Horse and Buggy, passenger mile—$2\frac{1}{2}$ cents

Horse and Buggy
EXPENSES CONTINUE
When not in use
Distance Traveled, 197 miles
COST

Hay	1.20
Oats	4.50
Straw	.30
Shoeing	.498
Grease	.0012
Depreciation	3.349
	$9.8482

Cost two passengers per mile   .05
Cost one passenger per mile   .025
Daily average distance 32 9-10 miles

To operate the Horse and Buggy the cost is based on hay at 1c. a pound, oats averaging 25c. a 4-quart feeding, purchased en route; straw at $1.50 cwt., 100 lbs. per mo.; shoeing, $2.50 a month; grease, 12c. a pound; depreciation, horse, buggy and harness, costing $375.00, .017 a mile, average 33 miles a day. Stabling not included.

THE ECONOMY CAR
"Perfectly Simple—Simply Perfect"
Maxwell Model Q-11, 4 cyl. 22 H. P.—$900

Automobile
EXPENSES STOP
When not in use
Model Q-11 Maxwell
Distance Traveled, 457 miles
COST

Gasoline	5.60
Oil	.60
Grease	.13
Depreciation, Car	3.66
Depreciation, Tires	6.85
	$16.84

Cost two passengers per mile   .037
Cost one passenger per mile   .018
Daily average distance, 76 3-10 miles

To operate the Automobile the cost is based on gasoline purchased at 16, 18 and 20c. per gallon, and oil at 65c. a gallon, purchased en route; grease at 12c. a pound; car depreciation, .008 per mile; tire depreciation, .015 per mile. Garage not included.

I've Proved Just What You Want to Know

What Prompted the Test Millions of people buy buggies yearly, believing them to be the most economical form of transportation. Thousands of farmers, merchants and professional men wrote me during the past year, saying "they would gladly purchase an automobile if they only knew it would serve their needs as economically as the horse and buggy."

I knew the Maxwell Model Q-11 Runabout, in its present state of perfection, could do three times as much work at half the cost and one-fourth the trouble, and decided to make a public test to prove it.

I Invited the Contest Board of the American Automobile Association to conduct this test, in order that it might be in disinterested control. The Board appointed judges to attest its results. The two vehicles ran each day over a predetermined route about the streets of New York City and suburbs. Each ran continuously for six hours, regarded as a normal day's work. Account was kept of every item of expense entailed. The needs of each vehicle were supplied at roadside stores at current market prices.

Each Day a Different Route was laid out, in order to cover all conditions of city and suburban traffic and all sorts of roads. One day they covered the densely congested districts of the city; another day they ran in infrequently traveled suburban roads. Everything was done to make the test normal, actual, eminently fair and conclusive, and the results confirm our claim that **Maxwell automobiles are cheaper to use than a horse and buggy.**

What It Means to You The Maxwell automobile means new life on the farm, happier and healthier conditions, better schooling for the children, by enabling them to attend town schools. Easy shopping for the wife. Quicker errands about the farm and increased knowledge of business conditions in your section of the country. It will, if properly employed, increase your earning power, save your time, extend your field of action and keep the boys on the farm by its money-making possibilities. It is always ready, sanitary, and cannot be overworked.

It Proved Beyond Dispute That the Maxwell Runabout is undoubtedly economical; that its pleasures are within the reach of men of moderate means; that it can travel three times as far, in the same time, as the horse and buggy, and, unlike the latter, can repeat the performance if found necessary; that its cost is about one-half; that it needs little or no care, as against constant care, and that while not in use expenses stop, while with the horse and buggy expenses never stop; that the Maxwell car is an efficient, economical, reliable, utility automobile, as near perfection as modern workmanship can make it.

These Books Free I want you to know more about this test. Let me send you all the evidence, also our latest catalogue and Farmers' Economy Booklet. I will gladly send you these books free, and the sending places you under no obligation. A postal will do. Just say, *"Mail Books."*

J. D. Maxwell   Pres. and Gen'l Mgr.

SALE OF MAXWELLS TO DATE

Sold to Sept. 30, 1910	37,389
Sold during Oct., 1910	1,767
Maxwells in use today	39,156

WATCH THE FIGURES GROW

MAXWELL-BRISCOE MOTOR CO.

Watt Street   **TARRYTOWN, NEW YORK**

BEGINNING OF THE END

efficient than horses in short hauls in the city.

"Within a radius of three miles we cannot use a truck at all as the expense is too great, one says.

"One brewery reports it can use trucks to advantage on long distance hauls". (*Breeder's Gazette*, Jan. 1, 1913)

"The horse has had to combat various scares since the time the old Erie Canal was dug and later steam railroads, then electric trains, bicycles, and now automobiles.

"There is now a scarcity of horses. $600 is a common price for pairs for business or farming purposes, but there are more horses in the country than ever.

"There will be a greater demand for coach horses and horses for driving purposes.

"The automobile and truck have come to stay for certain purposes.

"The real earnings of the power machines are a joke compared with their cost and upkeep." (*Breeder's Gazette*, Oct. 23, 1913)

Some companies, such as Consumer's (ice and coal) in Chicago, in 1914 used as many as 1,300 horses. They also had 50 trucks which they found economical for only long hauls.

Late in 1913 a group of carriage builders, wheelwrights, horse breeders, wagon companies, manufacturers of horseshoes, carriage lamps, etc. decided that in unity was strength. They formed the National Association of Allied Horse Interests.

Nevertheless, the move to the motor age continued on its inexorable route. By 1913, more and more motorized vehicles entered the scene. Ads appeared for the Haynes, Maxwell, Reo, Hupmobile, Jeffery and Studebaker, and for the Willys-Overland for $950. Farm tractors advertised included the

Leader, Heider, Hart-Parr and the Wadsworth Kerosene. A year later, advertising appeared for the Avery, Rumley Emerson and Bullock Creeping tractors. Auto ads in 1914 included the Pioneer Cycle Car, Cadillac, Lambert, Hudson, Jackson, Saxon, Mitchell and Velie. And trucks joined the takeover with advertising for Federals.

Finally, the *Breeder's Gazette* began to admit the inevitable, as seen in an editorial in their April 23, 1914 issue.

"Regardless of our personal feeling in the matter, it must be admitted that the truck has supplanted the draft horse very appreciably in the work of transferring heavy loads of the cities.

"However, in Boston the horse is holding its own.

"One is impressed by the number of two-, four-, even six-horse teams of glossy blacks, bays, and greys on every hand.

"Boston's narrow, crooked streets is one reason.

"We have only 150 horses now in the (Chicago) stables, said a spokesman for Armour & Co., as our use of the motor trucks has greatly reduced our need of horse labor. The horses we do have weigh between 1,300 and 2,000 pounds and most are high grades." (*Breeder's Gazette*, Oct. 29, 1914)

Other magazines began to see the handwriting on the wall. In his "Report of the National Carriage Builders' Association," President P.E. Ebrenz commented on the demand for horses for city use. "We will have to look the situation straight in the face and admit the automobile has hurt the demand for pleasure horses in city use."

The *Breeder's Gazette* on July 29, 1915, said, "Horses are getting dearer and trucks cheaper. Prices have more to do with dis-

THE END OF AN ERA. The horse era was drawing to a close as shown in this photo. In Milwaukee in 1912, carriages and limousines line up waiting for their owners who will soon emerge from a concert at the Pabst Theatre. There are ten horse-drawn rigs, and ten autos in the photo.

THE "OLD SETTLER'S PICNIC". This event was held every year at the same park in Union Grove, Wisconsin. The same parking lot was always used. These two photos, one taken in 1901 and the other in 1919 tell the whole story.

BEGINNING OF THE END

placement of horses by motors than any other consideration."

World War I probably put a damper on the acceleration of truck use in American cities. The draft horses presently in use were getting the chores done as economically as always. Trucks, tires, gas and parts were in short supply. These problems raised some false hopes.

"We are not likely in the near future to say goodbye horse, or to think we have reached the horseless age." (*The Blacksmith and Wheelwright,* Oct., 1915)

Wayne Dinsmore, Secretary of the Horse and Mule Association of America, said, "The draft horse still occupies the most important place by far in moving the commerce of the world and there is no reason to believe that he will ever cease to be a most important factor in city traffic."

"Chicago, today, has 32,759 one-horse vehicles; 13,488 two-horse vehicles; 212 three-horse vehicles and 11 four-horse vehicles." (*Breeder's Gazette*, Aug. 2, 1917)

In the same issue the Gazette editorialized, "Positive claim has been made all along by practical city transfer companies that they could not do their work without horses.

"Horses are so much more efficient in starting an overload, handling a load in soft alleys and around railway tracks and freight platforms and docks that they retain high esteem."

Just before the end of World War I, an article in the *Breeder's Gazette* of Sept. 19, 1918, was headlined "City Trade in Horses Improving." Excerpts reflect a positive horse market once again: "Pre-war conditions exist in the horse market.——Under present conditions an ill-conditioned horse is practically unsaleable. Quality is demanded——the day of the

merely serviceable horse is over.——Revival of city demand is noticeable. Department stores that professed to have discarded the one horse delivery wagon are reinstating it and orders for low-down 1,500-pound chunks at $200 have been filled.——A somewhat brisk demand for teams and single wagon stock weighing 1,450 pounds to 1,500 pounds at $200 to $225 has been developed."

The die was definitely cast, however, by 1918. Many farm tractors were available such as Cleveland, the E-B, Waterloo Boy, Moline, LaCrosse, IHC, Indiana, and the Rock Island. New advertisements showed up for Patriot trucks and Franklin, Paige, and Chandler autos.

Roads improved, streets were paved, mechanical vehicles got better and companies began in earnest the relentless switch away from horses, as did farmers, individuals, fire departments and other users of the draft horse.

As things began to get back to normal after the war, occasional items appeared that

gave the impression horses on city streets were becoming a nuisance.

"*The Chicago Tribune*, in an editorial of Aug. 28, 1919, said, 'The automobile tax is based on the fact that they wear out the roads

THE PASSING OF THE HORSE

THE buggy building industry was exceptionally good during 1913. Some have said that more than a million of them were turned out. One concern in Indiana has more than 80,000 to its credit and it takes a very rosy view of the future, expecting to build close to 100,000 jobs during 1914. Did you ever stop to think what a hundred thousand buggies would look like if they were assembled in one place? Of course, most of these buggies were cheap jobs because those who formerly bought the high priced rigs are now indulging their fancies in $3,000 to $5,000 automobiles.

The automobile business on the contrary has had its ups and downs, making its greater progress in cheaper lines. The country was somewhat startled at the beginning of the new year by the announcement that Mr. Ford would divide $10,000,000 during the coming year with his employees and some of the critics· of the horse were loud in their assertions that man's best friend was fast becoming a relic of the past. The Ford idea is, however, but a incident in the automobile industry and the failures in this line which have been frequent have not been confined entirely to the small fry.

Buggy men who know what the demand for their goods is seem full of confidence. Plans have been suggested in some instances to do some general advertising on the advantages of the horse drawn vehicle just as the automobile builders had the sense to do in relation to their products when the buggy builders were in the dumps and not making a move to help themselves. Sometime ago there was organized an association called the National Association of Allied Horse Interests and they too have done some very satisfactory work in bringing to the attention of the general public some of the silly claims of the automobile and truck manufacturers.

The harness trade have been talking of getting together and now and then we hear of a blacksmiths' and wagonmakers' convention and there seems to be some hope that all the interests concerned will eventually realize that people are not so closely wedded to the automobile industry as some of the manufacturers would have us believe.

Occasionally the automobile papers take a jab at the horse and predict his ultimate passing. It seems to have about as much effect upon the horse as some of the religious organizations who are continually predicting the end of the world. Both bring a laugh, and both will be accomplished about the same time.

Horses are just as much in demand today as they ever were and are bringing better prices. The horse has many uses that the automobile can never expect to supplant. Horse racing goes on each year developing a better class of animals and more enthusiasm on the part of the public.

The busy business man is finding that long country rides early in the morning. astride a fast stepping saddler are more exhilarating than the continued consumption of the automobile cocktail, composed of fifty parts road dust and an equal quantity of naseating gasoline oders. Then too, we cannot expect society women to be long contented in wearing dust proof clothing. The natural tendency of woman is to make herself attractive to man and as soon as the novelty of owning as good or better car than one's neighbors is worn off we may expect the society folk to go back to the slower horse drawn vehicle which affords an opportunity of a greater display of feminine finery.

No one disputes the utility of the horse when it comes to the actual performance of work. Farmers are going broke every day attempting to supplant the horse with the tractor and many of the big city firms are going back to horse drawn vehicles after having given the motor truck a thorough trial.

So, taking it all in all, the passing of the horse seems rather remote and we may expect that there will still be ample room for the buggy builder, the harness maker and the blacksmith.

THE "CROW BAR" magazine ran this editorial in their February 1914 issue.

Credit: Author's Collection

BEGINNING OF THE END

and justly should bear a share of the cost of upkeep. It would be equally just to put a tax on city horses which is one of the chief factors of street cleaning expense.' " (*Breeder's Gazette*, Sept. 18, 1919) This was a problem—and one horsemen couldn't sweep under the rug. American cities had millions of gallons of urine and tons of manure dumped on their streets.

In the Oct. 23, 1919, issue, the *Gazette* listed the New York City horse population, emphasizing the startling decline:

1910 — 128,224 horses
1917 — 108,036 horses
1919 — 75,740 horses

Seemingly quite worried about the decline in horses and transition to trucks, a solid group of people interested in oat-eating horse power came up with a new organization to combat the downhill slide.

"The Horse Publicity Association of America was incorporated for the purpose of advertising the horse as an efficient and economical source of power in the field of usefulness.

"The officers announced the horse does not need protection, it needs publicity.

"Those back of the new organization are confident that the horse will soon come into his own again when the people realize his strong position in the general economy of traffic and farm work.

"We do not seek to displace the motor industry, but it should not be allowed to under-

THE OLD AND THE NEW. Many a three-horse team could haul a load equal to the capacity of the semi-trailer, but, unfortunately, at too slow a speed. Yet one Milwaukee trucking concern stated they made more money with 25 teams than 25 trucks.

Teams and Trucks.

An inefficient horse is enough to discourage anyone. Far be it from me to endorse its use. An inefficient machine is enough to disgust the buyer. Good breeders have taken all guesswork out of the production of the efficient horse. They have said that with selection and discrimination in breeding the cost of raising is practically the same. There is no longer any qualification of the statement that "the best pays best." With the gradual perfection of mechanical power, the general stock of work horses has been somewhat lowered. The opposite should be the effect. Producers of horses, like the makers of the inefficient machine, should accept the challenge, examine their stock, analyze its needs, and then live up to them.

The heavy draft horse has not suffered particularly in the estimation of truckers, teamsters and cartage and transfer companies for the ordinary demands of the day's work. Nationally-known transportation users declare unqualifiedly in favor of horses on all short-haul work. One particular instance of head work has brought horse use for city hauling into the spotlight because of the economies resulting. The A. T. Willett Teaming Co. of Chicago through its secretary-treasurer says: "Every delivery problem I have ever studied has been 75 percent a problem of organization and only 25 percent a problem of transportation. Adopting the opertaing economies perfected by the railroads, the heavy trainload and the Mikado engine, we introduced to merchandise teaming in Chicago :. (1) 3,400-pound teams instead of 2,600-pound teams. (2) 10,000-pound ball-bearing double wagons instead of 6,000-pound wagons. (3) The wheels underneath wide bed wagons for light or bulky freight. (These wagons haul six pianos instead of the three on a regular wagon.) (4) The ball-bearing, three-horse wagon, loading normally 16,000 pounds to replace the old four-horse team which normally loaded 13,000 pounds. Through standardization of wagons, harness and equipment, together with shoeing at night, we have almost completely eliminated 'time out' during working hours. Our equipment is organized to function 100 percent efficient."

The Chapman Lumber. Co., Syracuse, N. Y., has experimented with all kinds and makes of motive power. It is now using horses only in its work. The treasurer states: "We have a good average lot of horses which are doing our hauling in a satisfactory manner, but we plan to replace these horses with the very best draft teams that money can buy. We recently purchased a 3,600-pound pair of matched grade Belgian mares for which we paid $1,200. After studying the matter carefully we bought these mares, not for show or advertising purposes, but as a cold, hard business investment. We consider them the best and most satisfactory power-hauling unit that we ever owned, and as fast as possible we shall change our entire stable and replace with the best draft geldings that we can buy."

The J. M. Horton Ice Cream Co. of New York uses 350 horses and 60 trucks. The horses cover all stops within a radius of 10 miles from the plant; the trucks are used on territory outside of that. The superintendent of delivery for the company says: "It is all bunk saying that motor trucks are more efficient than horses for city delivery. Our experience is all in favor of the horse. Our horse teams make only one trip a day. They cover between 15 and 20 miles and make from 35 to 40 stops. A motor truck cannot cover any more territory on these many-stop trips, and our heavy trucks will not make more than 3 miles on a gallon of gasoline. In the winter, when the snow is heavy, and it is necessary to keep the engine running at stops, frequently trucks do not average over 1 mile per gallon. Even under ordinary conditions our operation costs are all in favor of horses. We can run a one-horse rig for $10 per day, including driver's wages and all overhead. A two-horse team costs $15 per day, including a helper as well as the driver. It costs us $25 a day to operate a truck. We can do just as much work with a two-horse team as we can with a truck, and do it just as well. A truck costs us $10 per day more, without giving us any better service."

The Syracuse Ice Cream Co. uses trucks for all long-distance runs, but for city delivery work President Dennis says: "I am a little proud of our delivery horses, consisting of thirty-two dapple-gray grade Percherons, used on white wagons. This outfit makes a good advertisement, a well-known trademark for the Syracuse Ice Cream Co. in this city and surrounding country."

A good truck is a powerful time-saver on long runs. Its use on trips which are heavy and long has established its superiority for that class of work, just as the ease of handling and economy of the horse has won the short-haul for horses. The Consolidated Ice Co. of Pittsburgh states: "The horse is far more successful in the delivery of ice in congested and in resident districts, but for long and difficult hauls the motor vehicle of course has the preference."

The White Swan Laundry Co., Birmingham, Ala., says: "Of course on long, suburban routes we find automobiles more satisfactory, from a long-run standpoint, but they cost more." The Morrow Transfer & Storage Co., Atlanta, Ga., uses horses and mules for short-haul work, and adds, "It has also been our experience that on long hauls the truck can be used to greater economy than the horse or mule-drawn wagon."

If horses are necessarily kept for other work, and are free and available for long hauls, it obviously is better organization to use them on long hauls than to let them stand idle. Such is the usual farm situation. Horses on the farm for other work, with good management, usually can handle the hauling. Special exceptions to this are the fruit and garden truck farms, where a quick early morning market adds to the salability of the produce. Daily trips to a distant station will warrant an investment in a good truck. Other considerations are keeping horses on the farm as the main source of power. Their adaptability, in different combinations of units for all varieties of field work, and the fact that they are self-perpetuating, raised at cost, and salable before depreciation begins (if they are good, efficient animals), mean much to the businessman-farmer. He also counts in the contribution to soil fertility, which, at present prices, is worth from $66 to $88 per horse each year. On the farm, as in the city, good management and economical discrimination must decide. Here, as everywhere, the investor must make sure before he goes ahead, if he would avoid loss. Good efficient trucks and good efficient horses, fitted to their particular work, are a source of satisfaction and profit to their owners. Let each power unit be in its place; then neither will suffer at the expense of the other.—WAYNE DINSMORE, Secretary The Horse Association of America.

VARIOUS COMPANIES reported on their use of teams and trucks in the July 15, 1920, issue of the *Breeder's Gazette*.

 # BEGINNING OF THE END

mine the economic basis of our country. We have the facts to show that motor driven machines can never supplant the horse within a certain range of efficiency—up to 10 to 12 miles at least.

"We will show you the economic crime of building all the roads for motor traffic whereas the sensible plan would be to put a strip of highway designed for horse-drawn vehicles alongside every fancy motor road." (*Breeder's Gazette*, Oct. 30, 1919)

Economically, draft horses delivering the merchandise were cheaper on short runs but on longer routes of 10 miles or more horses could not do anything to speed up their average pace of three and one-half miles per hour at a walk.

"Wilson & Company, meat packers, came up with figures to give an idea of the comparison between delivery costs with trucks and with horses.

"The average trip during the period of the records was 10 miles.

"The cost of delivery the first year with a two-ton auto truck was 25¢ per 100 pounds and with a one-horse wagon 13.3¢. With a two-horse wagon 7.9¢ and with a four-horse wagon 5.9¢ (the more horses the bigger the wagon; thus the bigger the load hauled).

"The second year the cost was 30¢ for the auto truck, the increase being due to the depreciation and repairs; while the depreciation on the horse outfit was so small as not to affect the figures.

"It is stated that the motor outfit should last

WAR TIME SERVICE. Wagons were more difficult to locate than horses for companies that wanted to go that route during World War II.

THE LAST HURRAH. The last spurt in the use of draft horses came during World War II. Because of rationed vehicles, tires, and gasoline, companies realized the economy of horses, and resorted to them for some work. Here is one of the 35 teams Schlitz Brewery put to work hauling between freight depots and the plant.

FACELIFT. Motors and other unnecessary equipment were removed from sixteen laundry trucks in Hartford, Connecticut. The vehicles were then fitted with a pair of shafts and kept in service during the war. A supply of oats for a noon meal was carried in the tool compartment.

five years and the horse outfit 15 years. The initial cost of the two-ton auto truck was $2,200 and the two-horse wagon, harnesses, and horses $1,000.'' (*Breeder's Gazette*, Oct. 30, 1919)

Another bit of creeping evidence that the final days of using draft horses as the primary source of power were approaching was seen in the editorial in Nov. 27, 1919, *Breeder's Gazette*:

''Horse watering troughs are going out of fashion in the large cities. They have been condemned by sanitary boards and health officers.

''Horses require and deserve cool, pure drinking water frequently while at work, especially in hot weather.''

(About 15 years later the Wadhams Oil Co. of Milwaukee, as a well-intentioned and humane publicity stunt, ordered all of its gasoline service station attendants throughout the city to water any horses when drivers brought them in.)

In 1920 the Gazette reported that Chicago had only 40,000 work horses, down from 80,000 in 1910. However, the ominous figure to horsemen was that Chicago had 19,550 motor driven vehicles in 1920 up, startlingly, from the 1911 figure of 800.

The Horse and Mule Association reported that in 1920 there were about 20 million horses and about five and one-half million mules on American farms.

In addition, there were just over two million horses and mules at work in cities and various industries.

Some companies, for purely economic reasons, continued with horses. One such operation was the Hecker-Jones-Jewell Milling Co. of New York City which turned out 10,000 barrels of flour a day.

In the Jan. 13, 1921 issue of the *Breeder's Gazette* it was reported about this company,

OUT OF MOTHBALLS. A guide service in Washington, D.C. brought this outfit out of storage for use during the war years. The tourists seemed to enjoy the tour immensely.

Credit: Philip Weber Collection

BEGINNING
OF THE END

"We operate teams and wagons in a 10-mile zone, electric trucks in a 20-mile zone, and gas trucks in a 40-mile zone. Horse drawn vehicles are of a five-ton capacity.

"Our horse-drawn wagons are the most economical vehicle we operate; otherwise we would not continue to keep up this department."

By 1928 the American Railway Express operated only 7,710 wagons and sleds and had 6,941 horses.

This was an abrupt drop since 1918 when the Express Company owned 15,600 wagons and sleds and 18,420 horses. They said trucks were simply faster.

The depression of the early 1930s kept a lot of horses on the farm. Oats were about 15¢ a bushel in 1932 so it was economical to keep, raise, and use horses on the farm.

"By 1937," the Horse & Mule Association reported, "the prices for good farm chunks, four to six years old, 15.1 hands, weighing 1,400-1,600 pounds were $400 to $450 a pair at the Chicago market." Prices declined, however, in the next few years.

With the advent of World War II the demand for horses and mules for export increased. In America, because of gas and tire rationing, many companies put horses back to work. Most notable were the dairies, since horses were excellent for the start-and-stop neighborhood deliveries.

One Hartford, Connecticut, laundry removed the motors from 16 delivery trucks, added a pair of shafts for the horse and carried on.

A sightseeing company in Washington, D.C. put a horse-drawn rig back on the streets to the delight of patrons.

The Checker Cab Company in Boston resurrected a fleet of Broughams, Victorias and other similar horse-drawn vehicles to handle the short run customers from the depot.

Schlitz Brewery had about 35 teams at work, used for intra-plant deliveries such as bringing cartons from warehouse to bottling house.

Many other companies throughout the country used horses for city deliveries again during the war years, sort of a last gasp for the draft horses.

The bureaucratic government in Washington, while appreciating the major effort of many companies to conserve gasoline and tires by using draft horses, threw the usual curve. The uninformed bureaucrats expected the horses to practically go barefooted as they accomplished their work. "The War Production Board restricted the use of metals in horseshoes to 90% of the 1940 production—they also specified shoes can only go to farmers. City dairies and bakeries, etc. will be unable to get any. The Giant Grip Manufacturing Company of Oshkosh, Wisconsin, one of the oldest and largest producers of horseshoes, is getting orders from city after city where firms are switching from trucks to horses," reported *The Milwaukee Journal*.

World War II created the last spurt in the use of draft horses in city streets. When the Japanese surrendered, the dependable and economical draft horses did likewise.

The machine age had taken over. However, before the age of gas buggies and trucks and while they were gradually being developed, horses moved people, merchandise and freight.

The photographs in the chapters which follow are a record of how horses accomplished those important jobs.

CHECK THE OIL? Service stations responded to the requests of the few peddlers on the streets. When they said, "Fill 'er up," it was with water, not gasoline as this 1944 photo depicts.

CHAPTER II

TRAFFIC, STREETS AND ROADS

HORSES were versatile. They could navigate loads over any lane, path or road, whether hard surfaced, mud or sand. They could haul loads even where there was no road at all.

Early on, some towns began to pave their main streets with cobblestones. These naturally rounded stones, preferably somewhat flat sided and six to 10 inches in diameter, were easy to find. Thus, they were probably the first product for paving streets in communities. Cobblestone streets were used in the late 18th century through the next 150 years.

The first "engineered" road in America was a 62-mile stone and gravel stretch built in 1793 and 1794 between Philadelphia and Lancaster, Pennsylvania. The first highway built with Federal funds was the 72-mile Cumberland Road from the Potomac River to Cumberland, Maryland, in 1840.

During the first half of the 19th century, and a few years before, over 4,000 miles of canals were constructed in America. Enormous tonnage was floated on these ingeniously built water highways.

Mules were the primary source of power, although in certain periods and areas horses

TRAFFIC JAM. Motormen on the trolleys clang bells, and drivers curse, but no horns can be heard. This snarl of traffic occurred in Chicago in 1909 at the corner of Dearborn and Randolph Streets.

HAULING FREIGHT. Commerce is moving with horsepower, including beer, spring water, carpets, ice, barrels and general freight, on Broadway and Chambers Streets, New York City.

Credit: Motor Vehicle Manufacturer's Assn.

Credit: State Historical Society of Wisconsin

STREET FAIR. The main street in Princeton, Wisconsin, was crowded on monthly Fair Days. Bob sleds, carriages and wagons line the curbs, as blanketed horses wait patiently while their owners hob-nob, gossip, barter, buy, sell and exchange ideas on the price of crops and livestock. C-1908.

DIRECTING TRAFFIC. The mounted policeman is stopping wagons in this photo while pedestrians cross the street. Judging from the heavy coats everyone was wearing, it must have been a cold winter day. Some of the horses are equipped with protector pads to keep the raw wind off their chests.

Credit: Motor Vehicle Manufacturer's Assn.

TRAFFIC, STREETS AND ROADS

did the towing of the boats and barges.

Canal companies would either rent mules or own them outright and, in turn, lease them to the boatmen for their trips through the system. Mules were most satisfactory because they fit the job requirement for slow, plodding and steady animals. Horses were too high strung, in comparison, for the towpath work.

Surfaced roads between cities were slow in coming. By the mid-1840s wooden plank roads began to appear and in the next 10 years thousands of miles were built, most in areas where there was standing timber. They might have been relatively inexpensive and easy to maintain, but they were obviously vulnerable to rotting which shortened their useful life.

Nationally, with the big push on developing railroads in the 1830s, 1840s and beyond, highway construction slowed down.

MAIN STREET. Freighters lined the curbs and gave this Helena, Montana, street a different look than Eastern cities had in 1869. Various business signs mentioned dry goods, Justice of the Peace, saddlery, liquor, tailoring and a drug store.

HORSE-DRAWN COMMERCE. Chicago's South Water Street Market (in 1910) was jammed with vehicles loading, unloading and waiting a turn to back into the curb so the drivers could conduct business.

KEEPING THE STREETS CLEAN. The street cleaner with his cart was a regular sight on New York streets when this photo was taken in 1900. His main chore was to pick up manure—and there were tons dropped each day.

Credit: New York Public Library

(In a way this was a tribute to the draft horses, as they continued to haul enormous tonnage of freight wherever it was needed—paved roads or not.)

Bicycle riders became more prevalent in the 1880s and they demanded better roads since they found it impossible to get anywhere over mud, ruts, or sand.

Out in Ohio's clay regions, brick streets began to appear in the mid-1890s. Gravel and tar roads began to show up around 1906 and the first concrete country road was built three years later.

"Wide tires for wagons", was the headline of an article in *Breeder's Gazette* of July 6, 1892. "Heavy vehicles should have six-inch wide tires; front axles should be shorter so that the four wheels roll a portion of the road two feet wide at each passing. This would measurably improve roads. Narrow wheels cut up roads if the vehicle carries a heavy load.

"There should be a tax on a person using a tire narrower than 3 inches and the tax should be reduced by one-fourth when a four-inch tire is used.

"The draft of a wide-tired vehicle is not increased but diminished. Farmers are advised not to drive in ruts as this is destructive to the road. Drivers should not follow in the tracks of another wagon. Wagon makers should be made to show cause why the wide tire cannot be bought in most sections of the country except on special order."

A few weeks later Clem Studebaker answered. "On wide tires, the supply of timber for the felloes is less abundant. In case of an increased demand on the timber supply, already scarce, the tendency will be to still higher prices for wide-tired wagons."

By 1907 there were two million miles of roads in America but only 153,662 miles had any kind of surfacing. Some communities passed ordinances trying to control certain

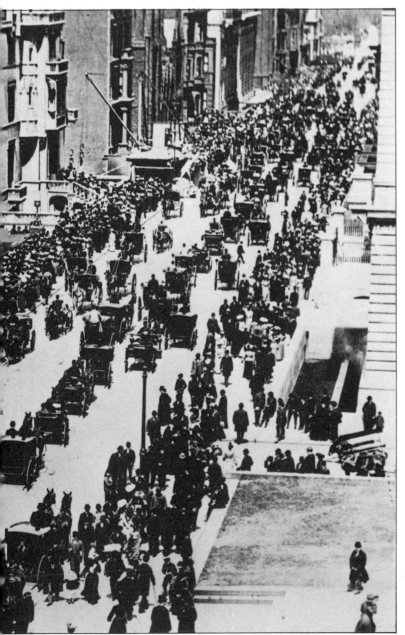

Credit: National Archives

EASTER MORNING, 1900. New York's Fifth Ave. was bustling with horse-drawn cabs and carriages. Two motor cars are included in the mix.

27

Credit: Gene Baxter Collection

TRAFFIC, STREETS AND ROADS

problems with horses and wagons: ● Hitching posts had to be at least three feet high. ● $1.00 fine if any vehicle was left on the street longer than two hours. ● Any wagon with a load of more than 4,000 pounds must have wheels with tires a minimum of three inches wide.

It was pointed out that "a 1,200 pound horse can pull a load of 2,000 pounds on a level dirt road but can haul 5,000 pounds as easily on a level macadam road in average conditions". (*Breeder's Gazette,* Dec. 4, 1912)

City streets were being surfaced with brick, granite blocks, packed gravel, packed sand and clay, creosote wood blocks, and—in a few instances—macadam and concrete. Motorists and cyclists loved this and it surely made the hauling jobs easier on the draft horses.

"Anyone living in a country town can tell just what condition the roads are in without asking questions. If many autos come to town the roads are good—if the hitching racks are thickly lined with teams the mud is deep. The auto is convenient and the horse is indispensable.

"Last summer and two years ago Cornbelt roads were dry and hard for months at a stretch. This summer with rain at the rate of 10 inches a month they realize the horse furnishes the only sure mode of country travel." (*Breeder's Gazette,* July 22, 1915)

In the same issue, the Gazette reported

WAIT YOUR TURN. The team on the creamery wagon in Troy, New York, is held up while a carriage passes in front. For this kind of traffic the granite block paving held up admirably. One of the buildings on the right has a poster proclaiming Ringling Bros. Circus will be in town Friday, May 29, which dates this photo as being taken in 1903.

When Horsepower Wasn't Under the Hood

In 1905 Milwaukee Had More Than 12,000 Horses, Boys Dreamed About High Stepping Trotters and It Was Illegal to Leave Old Dobbin Untied

By JAY SCRIBA, of The Journal Staff

A CENTURY ago there was no such thing as a snow plow in snowy New England. Instead, each town had a horse drawn snow roller which packed down the snow to make a firm track for the farmers' sled runners. Each town also had a snow warden who used horse drawn drags to pave a hard snow surface inside the covered bridges.

Like many vestiges of America's "age of the horse," the snow roller has become such an oddity that historical societies display it as a tourist puzzler. Yet scarcely a generation has passed since the average white collar Milwaukeean knew more about horses than the average farmer does today.

Until World War I, and into the 1920's, every teen age boy knew that you dasn't give "Prince" or "Buster" his bit on a zero morning without first blowing on it and warming it in your hands — otherwise the frosty metal would tear skin from his mouth. If it was really cold, you might fit his sensitive ears with a pair of woolen ear mitts crocheted by your mother. If it was slippery, you joined the long line of buggies outside the blacksmith shop, where hairy chested men in leather aprons reshod horses with steel calked shoes, or shoes nubbed with the new fangled "borium" alloy, "so hard that it would scratch a railroad track."

Kick Could Result From Carelessness

You knew that the first thing to do with a mean "skate" was put on a "blinder" halter, as a horse's eye focuses slower than a human's, and any unidentified object approaching suddenly from behind is likely to be kicked into the manure pile. Just as today's young men talk about "mag" wheels and "spoilers", you knew all about such things as surcingles, wiffletrees, hames, cruppers, hocks, withers, pasterns, stallion shields, nose bags and moon blindness.

Your traveling man uncle knew that even in a "dry" town, there was always a brown bottle buried in the oat bin at the livery stable. (Finding the livery stable in a strange town was simple—you just stepped off the train and followed your nose.) Conversely, your mother recognized the livery stable as a sink of iniquity second only to the pool hall—a fly ridden, tobacco splashed, carbolic acid reeking hangout for "rounders" who pored over the shocking pink pages of the Police Gazette. (This despite the fact that prominent male citizens dropped by to play checkers, pitch horseshoes or borrow an army blanket to sleep in the haymow when it was too dangerous to go home.)

Side Saddles Popular in Central Park

Your father prized his shiny rattan buggy whip, with its mother of pearl handle, and argued with his cronies about whether a 2,000 pound brewery Percheron really ate a 65 pound bale of timothy every day. Your older sister loved horses too, but was not about to ride one, not even aboard one of those fancy side saddles said to be the rage in Central park. She had read in a health book that, for ladies, horseback riding produces "an unnatural consolidation of the bones of the lower body, ensuring a frightful impediment to future functions which need not here be dwelt upon."

In 1905 Milwaukee citizens owned more than 12,000 horses. In 1918, the United States horse population reached its peak of 21,550,000 animals. The city of Milwaukee owned fire horses until 1928, and police traffic horses—testy Socialist mayor Dan Hoan damned their riders as "Cossacks"—until 1949. Milwaukee breweries revived the use of brewery horses during World War II's gas shortage. The last work horses—used to pull trash wagons and peddlers' carts—departed for the mink farms and glue factories about 1963.

Today (except for a couple of hundred saddle horses) a Milwaukeean has to look sharp to find vestiges of the city's horse age. The most obvious holdovers are the many back yard carriage houses remodeled into apartments. Others include occasional hitching blocks, water troughs — most of them reduced to flower planters—and "step" blocks set near the curb to help long skirted ladies alight gracefully from carriages. The porte cochere roofs attached to grand old houses stand high over modern autos because they were built for tall carriage tops. The close grooving seen on old driveways and garage floors was intended to give calked horse shoes a purchase. A horse head emblem over a big door is almost certain evidence of an old livery stable.

Similarly, old country roads wind a lot because they were laid out to save horse and driver from problems with hills and swampy mud holes. The typical small town cemetery is a short buggy ride from town. Country roads that jog at bridges usually do so because the bridge was built next to an old ford, which marked the original roadway. Old roads on long hills were often built in a series of stepped plateaus, so as to give tired teams level resting places.

Sounds and Smells of the Horse Era

In the day of the horse, cities smelled rural and "horsey" from all the hay, grain and droppings. Loaded haywagons were a common sight in residential areas. So were wagons driven by grocers, icemen, milkmen, trash collectors and peddlers of all kinds. You could hear horses stomping in the shade to shake off flies, horses whinnying to each other across back fences. (A whinnying stallion on Wisconsin av. could send harnessed mares skittering and cause an awesome jam of interlocked buggy wheels and snarled harness.) You could hear wagon wheels booming on plank bridges, steel "tires" grinding on granite cobbles, hooves plop-plopping on cedar paving blocks.

The Milwaukee carriage house had a hayloft on the second floor and often rooms for grooms and footmen (who stood in liveried ranks along the fancy stores on E. Wisconsin av., waiting while madame shopped.) There were several stalls, heaped with straw or chemical smelling tanbark bought from the tanneries. Each stall had a salt block, wired to the wall above the feed bin, and a hoof pick for digging out the pebbles that could make a horse lame. Each had a rack of bottles full of murky elixers for curing spavin, croup, founder, pumice sole and navicularthritis, a dread ill which "afflicted the hinge between the coffin and coronet of a horse's foot."

There was a big chest for horse blankets and the buffalo robes used to keep warm in winter. A buffalo robe came from a real live buffalo, and looked as hairy and wild as if it had been peeled from a mastodon. The idea with a buffalo robe was that you put it over your lap and warmed the inside with a barn lantern or hot bricks wrapped in old carpet.

Sleigh Bells Were Standard Equipment

There was usually a cracked McClellan saddle hanging on the wall from the days when your grandfather wore his Union army uniform and rode with the Milwaukee Light Horse Squadron in the Fourth of July parade. There was always a leather strap of big brass sleigh bells, as a city ordinance required all sleighs to carry bells. (Other laws set fines for letting your horse nibble the bark off your neighbor's maples and for leaving a horse untied. The latter was something like leaving a car with its motor running, and was serious as it could lead to a dangerous runaway.)

The smallest child learned respect for horses, often after getting a stinging swat in the face from a switching tail. Another thing you remembered was your father setting you high on the broad back of the family's gentle Morgan, then chuckling at your clutching panic when the horse shook its skin—and you with it—to shake a fly.

In those days a 10 year old boy wanted nothing more from God or man than a high stepping trotter and a fancy buggy. Such a "coachy" rig would have varnished pull shafts and rubber tired wheels with shiny brass inner rims and either red or "natural" spokes. The hubs and all metal parts were nickled, except for the whip socket, which glittered richly in solid "German silver." The back of the seat had a wicker shoulder rest. The dashboard, which screened off mud and other objectionable things, was covered with black patent leather. At its corners were brass kerosene lamps, with big red rubies in back.

Price of Fine Rig Wasn't Hay

The tan harness had lots of brass balls, ivory rings and rubies let into rosettes at the horse's ears. As final touches, you could braid your horse's mane, tuck in a couple of red roses and buy a Dalmatian trained to trot elegantly underneath you between the wheels. You were now ready to "bomb" Lake dr. or the avenue with the hottest sports in town.

The price of such a rig could equal that of an expensive car today. A farm hand might scrape for three or four years to buy one for courting or the Sunday evening buggy race. After the honeymoon, he usually traded it in for a plain black buggy and a "general service horse." When his family increased, he traded up for a surrey, a four place vehicle which came with or without a fringe on top.

A buggy's side curtains always leaked rain, but could be buttoned fairly snug in winter. A writer recalls that a buggy had "a greater sense of security than any car. The isinglass peek hole might be covered with snow, but it made no difference. No wiper was necessary except on a road strange to the horse. A thump on the front curtain would usually take care of matters. There was an intimacy about buggy walls, too. Even heated with hot soapstones and heaped with feather ticks, a buggy was no traveling sofa. It was a place of protection where the elements were more markedly thwarted because of their imminence."

MILWAUKEE'S HORSE-DRAWN ERA. Jay Scriba of *The Milwaukee Journal* wrote this entertaining article in 1967.

READY FOR RUNNERS. The snow roller was used in many towns in lieu of plowing. The roller packed down the snow, making the streets suitable for sleighs. Snow wardens patrolled the streets to be sure snow was put on any areas blown bare by wind. This photo was taken in Boonville, New York.

TRAFFIC, STREETS AND ROADS

that the Ohio State Highway Department planned to have all cities connected by hard roads ample to take care of all traffic any season of the year. This kind of progress and planning varied from town to town, county to county, and state to state. Some were more progressive than others, some had a larger population and thus, in turn, a better tax base with which to develop the roads.

By 1916 the U.S. Congress passed a Federal Highway Act appropriating $75 million for interstate highways. In the next eight or nine years, 50,000 miles of roads paved with brick, gravel and concrete were constructed.

PROMOTIONAL PARADE. McCormick day in Fountain City, Wisconsin, around 1890, was bright, sunny and warm. Plenty of barefooted kids were on hand. (The rider of the mule in the lower left corner finds the hot day too dry for his taste!) The parade seems about to start—most wagons carried a load of McCormick farm machinery parts, except the vehicle on the left that's loaded with musicians. It will be a fine promotion for the McCormick dealer as the parade wends its way through town over dirt streets.

OIL BOOM. As an oil boom town, Seminole, Oklahoma, was too busy to do much about the streets, as evident in this photo taken about 1920.

BUILDING ROADS. Gangs of horse- and mule-drawn scrapers move the fill to a new grade in this road building scene in Milwaukee at the turn of the century.

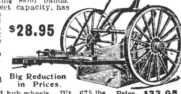

Credit: Author's Collection

FROM THEIR 1908 CATALOG. Sears, Roebuck offered these road scrapers in 9, 13 and 17 cubic foot sizes. Note Sears' assertion that they handle only one quality of scrapers—THE BEST.

EARTH MOVERS. These teams are moving earth on an Iowa road project in 1902. When the team gets the load to the right spot, the driver lifts the handles and dumps the dirt.

Credit: Caterpillar Tractor Co.

ROAD SCRAPERS IN ACTION. Bucket scrapers such as these usually came in three sizes: 3-1/2 cubic feet, 5 cubic feet, and 7 cubic feet.

Credit: Asphalt Institute

TEAMWORK. The span of mules on the right moved ahead with the big road scraper powered by eight mules ahead and four behind. One man on the scraper kept blade adjusted, while the other tended the conveyor belt.

Credit: State Historical Society of Wisconsin

Credit: Caterpillar Tractor Co.

MACHINES HELP. A Best Tractor powers the dirt-elevating grader loader while dump wagons receive their loads in the road-building operation pictured in this 1920 photo.

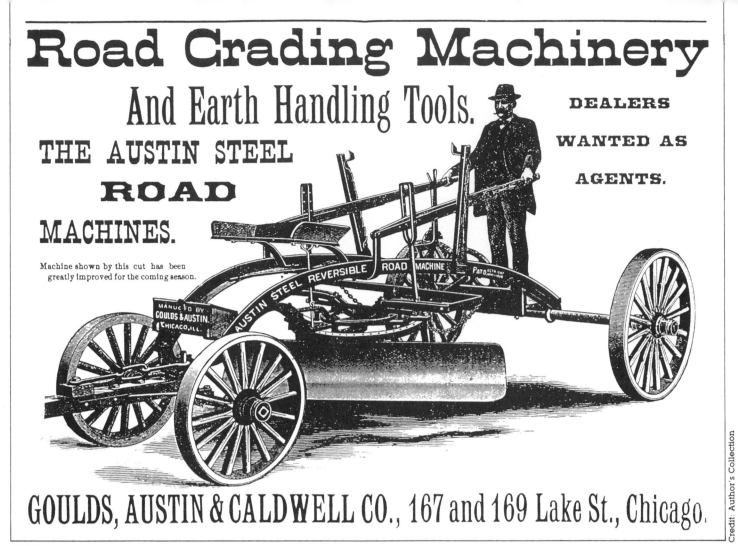

Road Grading Machinery

And Earth Handling Tools.

THE AUSTIN STEEL ROAD MACHINES.

DEALERS WANTED AS AGENTS.

Machine shown by this cut has been greatly improved for the coming season.

GOULDS, AUSTIN & CALDWELL CO., 167 and 169 Lake St., Chicago.

AN ADVERTISEMENT from 1889 for a road grading machine.

TRIMMING THE EDGES. This blade grader kept the weeds and grass off the part of the road that did not receive much traffic in 1910.

OILING DALE AVE. in Savannah, Georgia, preparatory to stoning around 1908.

THIS OIL TANK, which could carry a load of 275 gallons, was used by the Armour Packing Co. Note seat on the rear for the operator of the various valves and levers.

HEAVY HAULING. An enormous load of paving bricks being hauled in a Streich Dump wagon in 1920.

IMP. CUIRASSIER (44407)25574.

LONG-LASTING BUT BUMPY. Cobblestone streets were durable for horse-drawn traffic, but rough on the passengers. This street was laid in 1780 in Alexandria, Virginia.

PAVED WITH STONE. Granite blocks smoothed out the streets and never seemed to wear out.

FOLLOW THE RED BRICK ROAD. Red brick pavement was popular in areas where clay and brickyards were abundant.

THE WATER WAGON settled the dust on unpaved streets.

GETTING A REFILL. The tank probably held 500 gallons of water. A wide swath could be sprinkled with this rig.

STREET WASHER. Milwaukee had ten of these squeegee street cleaners to use around 1908 to 1914. They were made locally by the Kindling Machinery Co. The street department in Houston, Texas, said, "It wastes no water, creates no dust, sweeps clean, and the revolving rubber squeegee washes the pavement so completely that it leaves no slime, mud, or dust in its wake."

NO TRACTION. The horses are tense as they struggle up the street with their load sensing the lack of a secure grip underfoot. This team, coming up a hill with a heavy wagon, was in trouble.

THE BLACK is finally down but apparently uninjured.

THE DRIVER has sensibly unhooked the trace chains from the single-tree, while the horse struggles to retain its feet. The lack of caulks on the shoes is apparent.

A VANISHING SIGHT. When this photo was taken in Milwaukee in the 1930s, few horses were left on the streets. Here a team pulls a wagon while the crew spread a mix of sand and salt on the pavement at intersections.

HORSE-DRAWN SNOWPLOW. Scrapers were used to clear off snow from the sidewalks around parks and city-owned institutions in Milwaukee in 1944.

READY FOR A SLEIGH RIDE? Blanketed horses await their masters. Across the street a 4-horse team is leveling out the snowbank, spreading the snow out over the bare street—better for the sleigh traffic.

HELP FROM HORSES. The *Illustrated News* ran this picture in January, 1853, to show how heavy draft horses solved snow problems for the railroad.

INTO THE RIVER. Where to dump the snow that was shovelled off the streets? Milwaukee solved that problem by dumping it in the river. A plank was removed and the team straddled the opening. When the wagon was in position the driver stopped the horses, dumped the load and drove off. A workman stood by to shovel the excess into the opening, guide traffic around it and wait for the next load.

IN DEEP TROUBLE. Roads became quagmires when it rained. They were rutted and rough roads when dry. Any effort to sharply turn this buggy would put a severe strain on tongue and wheels.

AND THERE WERE MISHAPS. An axle nut fell off and this driver lost a rear wheel in Milwaukee around the turn of the century.

RURAL HIGHWAYS

Poor roads put the family "in a rut" and keep it there.

Good roads mean opportunity for

1. Neighborhood social life
2. Consolidated schools
3. Prompt mail service
4. Church attendance
5. Prompt medical attendance
6. Cheaper hauling of produce

THE PUSH FOR BETTER ROADS. Farmers were inclined to buck being taxed for road improvements which they felt would benefit the city population. Posters like this helped change rural attitude toward state-financed improved roads.

What Bad Roads Cost.

James R. Marker, Highway Commissioner of the State of Ohio, says that bad roads cost the country $7,500,000,000 a year.

"Enough money to build fifteen Panama Canals," he says, "would be saved each year in this country if we had good roads. We have figured that the cost of hauling one ton one mile on a good, hard, level road, by horse-drawn wagons, is 8 cents. The average cost in the United States is 23 cents a mile, and in certain sections of the country as high as 54 cents.

"Every year 5 billion tons of freight are hauled over the roads of the country and since the average haul is about 10 miles, the traffic amounts to 50 billion ton-miles a year. At the average of 23 cents a ton-mile for the entire country the cost of hauling by wagons reaches the staggering sum of 11 billion dollars a year. If this would only cost 8 cents a mile we would save 7½ billion dollars."

A TIGHT SPOT. An American Railway Express wagon was moving through an alley in Milwaukee in 1924 and the driver was guiding the team around an excavation when the off-horse shied and pushed his mate directly into the hole head first. It took the fire department, with the use of a crane, an hour to winch the animal out of the hole. The story in the paper said the horse would probably lose the sight of one eye.

STRANGE MISHAP. This accident occurred in Grand Rapids, Michigan, in the early 1920s. The draft horse pulling a wagon galloped out of control down a steep hill and crashed through the top of this parked automobile.

THIS ARTICLE appeared in the August 1914 issue of *The Blacksmith and Wheelwright*.

IN A RUT. This rut was so deep the buggy tipped over. One man holds the horse while the other tries to right the vehicle.

CHAPTER III

DELIVERING THE MERCHANDISE

COMPANIES delivering merchandise to residential areas in any city had a great sense of pride in the equipment with their name on it and in the looks of the horses that pulled their rigs. This held true whether the wagons were delivering beer, express, milk, bakery, liquor, meat, groceries or items purchased from department stores.

Horses that worked in the cities had more to concern them than just moving loads. They had to be trained to live with the din, hubbub, clatter and clamor of the environment.

In the Aug. 13, 1914 issue of *Breeder's Gazette,* the situation was explained in the article ''City-Broke Horses:''

''Nothing equals a city education for a horse. He comes off the farm full of life and

Credit: Richard Mueller Collection

MAKING MUSIC. It's hard to forget the sound of iron-tired milk wagon wheels squeaking through the dry snow when the temperature was zero or below.

A FRESH LOOK. Milk delivery wagons were always neat, clean, well-painted and gave an impression of freshness. Bowman operated in Chicago.

DELIVERING THE MERCHANDISE

vim, innocent of the sights and sounds which crowd the city's day from start to finish. On city streets he finds the experience wholly new, strange and affrighting. Hitched with a staid and seasoned mate, he enters the congested downtown traffic and is promptly engulfed in a maelstrom of currents which threaten to extinguish him. With a roar of the elevated railway overhead, mingle the clatter and clank of the trolley car, the startling exhaust and siren of the motor car, the clanging gong of the fire patrol, the madly galloping horses drawing the fire engines, and shouts and curses of teamsters with heavily laden wagons. It is indeed a poor place for colts, but with firm and kindly guidance the green horse quickly seasons to these sights and sounds until none of them moves him. He is city broke.

ALL YOUR GROCERY NEEDS. The wholesale grocer would deliver to the corner stores, hotels or restaurants. This Troy, New York, wholesaler advertised in 1890 that he had flour, tea, coffee, spices, vinegar, salt, syrups, butter, cheese and eggs.

STOCKING UP. Commission Row in Milwaukee, where the grocers picked up their requirements. One man is busy scraping up manure from the street.

PRODUCE 'TRUCKS'. The Public Market at Liberty St. and 4th St. in Troy, New York, was a busy place in 1910. The horses lined up head to head leaving an aisle behind the wagons for the patrons.

"And yet a big, upstanding, high headed, 1,700 pound horse stood hitched to a single lumber wagon on a west side street in Chicago. A trolley car rumbled past, the motorman sounding the gong and the horse merely stood at attention. A big seven passenger touring car painted a distinctive color and going as fast as the law allowed, approached and suddenly the horse seemed instinct with fright. With high held head, ears forward, rolling eyes and dilated nostrils he seemed just ready to jump out of his skin. The motor car flashed by and the horse did not bat an eyelash, but held his fascinated frightened gaze straight ahead on a yoke of oxen drawing a canvas covered prairie schooner. To him it was an affrighting sight—he was city broke."

The following excerpts are from a story which was written by E.T. Robbins, Associate Editor of the *Breeder's Gazette*. It appeared in the June 12, 1912 issue.

Delivery Horses in High Class City Service

AMONG the most attractive sights to be seen in a great city are the delivery outfits of the large retail stores. The beautifully molded, proudly standing horse, sporting a shimmering coat and shining harness, is impressive enough to catch the gaze even of the wealthy observer whose vision is accustomed to the elegant and costly. But this is only part of the picture. The driver in uniform, and the artistic lines and rich finish and furnishing of the rubber-tired delivery wagon complete a pleasing harmony of serviceable substance, cor-

DELIVERING
THE MERCHANDISE

rect outlines and blended and contrasted colors. Any woman is proud to have such a vehicle stand before her door while the active courteous delivery boy presents the package she purchased a few hours before. When the boy mounts the seat beside the driver and the equipage moves majestically along the boulevard it acquires added charm. The lofty head, the elastic step, the whirling wheels, the play of lights on the shining surfaces, illumine the picture with life; and it is the exultant life of plenty and prosperity. It is difficult to imagine a more alluring display of style.

It is the opinion of L.L. Timmons, who has had charge of the Chicago delivery service of Mandel Bros. for 20 years, that the horses have contributed more real advertising benefit to the business of the firm than any other line of expenditure. It is the frequent forceful and favorable presentation of a merchant's

MADE IN THE SHADE. This grocery wagon was equipped with an umbrella to protect the driver from sun and rain.

LIGHT LOAD. A neat little vehicle, that never had to carry a load of any weight.

HEAVY DUTY WAGON. Borden's, with their great variety of products, used wagons such as this for hauling to wholesalers and freight to the depots.

MEAT WAGON. Armour's name appeared as a brass trim on the harness.
This sturdy vehicle delivered meat to the butcher shops.

WITH A GERMAN FLAVOR. Mouth-watering products were delivered to restaurants and stores.

DELIVERING THE MERCHANDISE

ILLUSTRATED WAGON. Built around 1886, this flashy delivery wagon was decorated with a different scene on the opposite side. Note the elaborate scroll work.

name which impresses it indelibly upon the mind in intimate association with the best of goods and service. The delivery outfit which presents a perfect picture of solidity, luxury and good taste does that. A woman is strongly attracted to the store which promises a beautiful setting for the delivery of her purchase, and when the horse and wagon have fulfilled their mission and disappeared from view, she cherishes the remembrance with a satisfaction colored alike by the thoughts that she has bought the best of goods and that her neighbors are sure to know it.

Speak to a dealer of the Mandel delivery horse, and he will say, "Oh, yes, that is the best there is." Size is essential both for strength to handle the heavy vehicle at a trot and to combine with it fittingly in substantial appearance. The horse must have a neat head, long neck and long sloping shoulder to give lofty carriage and gayety of step. The chest must be deep to give constitution and capacity, but not wide enough to interfere with action. The step must be high and straight, and the horse must show a joy in going. In a word, the Mandel horse combines service with style; it is a big carriage horse.

Many a horse has been bought from the

ICE CREAM ON THE WAY. Compartments at each end of this wagon are refrigerated units cooled with ice which protects the ice cream on its way to customers. Note the platform steps on either side of the wagon.

HEAVY LOAD. A heavily constructed wagon such as this can carry a lot of weight, thus the four-horse team.

48

YES, WE HAVE BANANAS! This wagon hauled bananas from the refrigerated railroad cars to Commission Row in 1915.

MILK FOR MILWAUKEE. A large wholesale wagon makes the rounds of restaurants and large grocery stores in downtown Milwaukee. The empty milk bottle cases are stored on a rear platform.

CHICKENS STACKED HIGH. Live poultry is waiting to be unloaded.

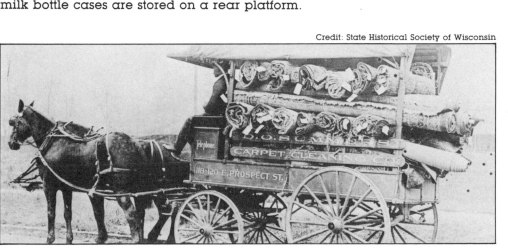

CARPET CLEANER'S CART. A hefty load was moved by this team.

DELIVERING THE MERCHANDISE

delivery wagon to fill the shafts of a wealthy lady's coupe. And why not? She picked the best-looking, proudest-going horse that she saw step past her door. He might have cost only $150 or $175 on the market, where only the wizard's eye could detect the diamond in the rough, but she paid $300 to $400 and in some cases even $700 for the privilege of calling him her own. For this reason it has been hard to keep the very best of the horses long in delivery service. The public has appreciated them too well.

All of the horses whose likenesses are here presented have been in the delivery service from 6 to twelve years. None is bought at less than 5 years, and 6 is preferred, because at that age the bones and joints are hard and not so liable subsequently to go unsound under the all-day hammering on the pavements. How well these horses have worn their clean legs, shapely feet and rotund forms reveal. They still work every day, going as

LUMBER CART. From railroad car to stack, from stack to shed, from shed to delivery, the horses moved lumber at the A.C. Beck Box and Lumber Co.

SUGAR SACKS. Two hundred sacks of sugar each weighing 100 lbs. makes a 20,000-lb. load. This team hauled the sugar from railroad yards to warehouse. The driver, Albert Butter, who had been driving horses for 50 years said he had 18 tons on this wagon on one occasion.

HAULING WOOD. In this 1915 photo taken in Evanston, Illinois, a load of lumber and shingles is headed out to a construction site.

HEADED FOR THE MILL. Huge bags of raw cotton would arrive in New Bedford, Massachusetts by railroad in the early 1900s. From the railroad yards, teams and wagons such as this would take the bags to the mill for processing.

IN A PICKLE. Farmers line up at a Heinz Pickle Plant waiting to unload the cucumbers. Wagons of all descriptions, and even a carriage, bring the produce to the packing plant.

DELIVERING THE MERCHANDISE

much as 20 miles and sometimes over 30 at a trot. Most of them work singly. They are the best-preserved old horses in such strenuous service in Chicago. Selection for their special work, and constant care in their management are responsible for their continued usefulness. No drunken drivers are trusted with these horses. The men are taught to start out at a moderate gait in the morning, and drive steadily rather than with bursts of speed.

One typical Mandel delivery horse is 16.1 hands high and weighs about 1,400 pounds. He is the best-looking old horse in the stable. Twelve years ago, when he had been only a short time in the delivery service, he was shown at the first horse show at the Coliseum. Although new to such business he stepped proudly around the arena, leading the delivery class with the rhythmical rush of a machine and the splendor of an aristocrat. He won the blue. He is not a high-strung horse, nor the highest-stepping but he is well made and has worn well.

The proudest old horse in the stable is also a prizewinner, and stands 16.3 hands. His shape was never the most attractive, but a

RECYCLER. Milwaukee Western Barrel Co. recycled used containers. The sign on the wagon said, "Buyers of Barrels and Butter Tubs". It is a bulky load but light weight, so one horse could handle it.

SHORT-HAUL. Some warehouses would hire transfer companies to take loads from the railroad yards to warehouses.

TRANSFER COMPANY. This company primarily hauled merchandise to and from railroad freight depots. Here are two well-painted and well-horsed rigs.

ONLY THE BEST. H.J. Heinz, founder of his famous company, had a deep interest in the horses pulling his wagons. They had the best stables, the best feed and the best of care. This team shows this attention. Most American companies did likewise.

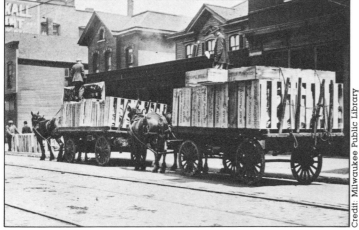

CRATED MOTORCYCLES. Coffee & Larkin Transfer Co. of Milwaukee (1914) have two loads of Indian motorcycles to haul to the dealer.

BAGS OF WHEAT. This photo appeared in a 1911 *Breeder's Gazette* and the caption said, "21,660 lbs. of wheat and wagon hauled by a pair of 1,350-lb. horses."

CORPORATE IMAGE. Companies had pride in their horses and vehicles. Heinz usually had black horses on their white wagons, as seen here. Armour used greys on yellow wagons. Other companies would pick other breeds or colors to complement their wagons. The Cream Maid Bakery in Hammond, Indiana, operated 50 wagons. Charles Dillner had the contract to furnish the 1,500 lb. dapple grey Percherons, all with white manes and tails.

DELIVERING
THE MERCHANDISE

courage and spirit have combined to give him a style and majesty of step all his own. Whenever he passes down the street people stop and look. If a horse ever has personality, he has it. No whip is needed to make him "show off"; he is always in exhibition mood. He takes the same pleasure in parading down the street now that he did when he began ten years ago. Although possessing a strong constitution, he is too high-strung to carry as much flesh as one would like. Like many another grand horse he looks much better in harness than out of it; and there he

TAKING A REST. While the bagged clover and alfalfa seed is being unloaded at the freight depot, the team stands hip shot, resting before heading back to the plant for another load.

54

CASES OF VEGETABLES. A load of canned goods like this at a Libby, McNeill & Libby plant required a well-constructed wagon and a six-horse team.

MAIL ORDER. Headed for the freight house with a load of catalog merchandise in 1906.

UNEEDA BISCUIT—OF COURSE!
The National Biscuit Co. used an enclosed van to protect their loads from the weather. Canvas could be rolled down to cover the open areas.

FULL WAGON. A 4-horse team heads for the freight house with a hefty load of merchandise.

EASY HAUL. A lightweight and bulky load enroute to the depot.

FANCY STYLING. A tobacconist's wagon was constructed in a novel way in 1885 by Abbot-Downing Co.

FLORIST'S RIG. Just about every company, big or small, had to have a vehicle of some kind.

LIGHT RIG. Printers needed one-horse vehicles for delivering small orders, picking up mail and doing other errands.

BULKY BUT LIGHTWEIGHT. A voluminous load, but light enough for one horse. This was Jac. Groeschel Tin & Furnace Works' wagon.

CHAIRS FOR SALE. In the late 1890s, this rocking chair peddler pulled a bulky load of probably homemade products. In summer months flies tormented horses; hence the fly net and ear caps. Some enterprising merchant had his name imprinted on neck covers and most likely gave them away as advertising.

BRAND NEW. This order wagon is fresh out of the shop in this 1908 photo.

DELIVERING THE MERCHANDISE

feels at home and glories in the chance to show what he can do.

He is evidently an overgrown standard-bred, and so is his mate. The three at the other end of the line have a dash of Percheron blood. The grays show it emphatically. They have a wonderful capacity for preserving a wealth of flesh and rotundity of form after 6 or 8 years of steady service, while their legs are as clean and fresh as when they were bought as 6-year-olds. They have proved to be very serviceable, although having less brilliant action than the horses of lighter build. Most of the horses in the stables appear to have a combination of trotting and Percheron blood probably being generally as much as three-fourths standard-bred. Obviously such horses are not commonly produced. Few straight standard-breds have the requisite substance, and few of more than one-fourth draft blood have the step and style.

It is scarcely hoped that the automobile will prove so attractive to customers. No machine can equal the horse in commanding attention. This fact is daily demonstrated in the downtown section where the veteran

high-stepper and his mate are used to deliver packages to guests at the best hotels. Limousines, taxicabs, touring cars and runabouts of the most luxurious patterns attract little comment in front of these institutions which house the wealthiest travelers of the land, but when the famous pair swings up to the curb there are always expressions of admiration. There can be no doubt that the "human interest" which is counted as a priceless asset in any modern enterprise is thoroughly aroused by the high-class delivery horse appropriately appointed.

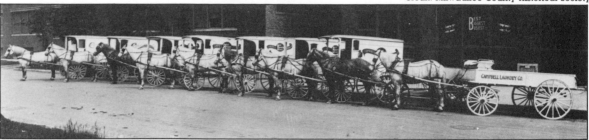

DAPPLE GREYS. The Campbell Laundry Co. of Milwaukee always used dapple grey Percherons on their white wagons.

A NEAT-LOOKING RIG. The harness appears to be nickle-plated.

RECYCLING CENTER. Peddlers, such as this one, picked up rags, paper, iron and other assorted junk that had value in various neighborhoods and then sold it to dealers at a small profit. It was estimated in the 1930s that approximately 200,000 junk dealers collected eight million tons of scrap iron a year.

WHERE IS HE? A hitching weight keeps the horse from wandering off. The horse is keeping an eye on the peddler.

LUNCH BREAK. Before the driver stepped into the coffee shop for his lunch, he put the nosebag of oats on his faithful horse. The canvas blanket kept the snow off the animal.

TRAVELING SALESMAN. The Watkins salesman, L.C. Smith of Oshkosh, Wisconsin, sold his Remedies and Extracts from house to house. Records indicate he frequently made up to $600.00 a day.

MILK BY MULE. Mules hauled milk on occasion, as is seen in this trim-looking span.

LOOKING GOOD. Dairies took pride in their horses and equipment —this Milwaukee dairy was no exception. Note the nosebag of oats hanging on the side door handle. This horse would not finish its morning route in time to get back to the barn by noon.

Frequently the milkman, if not hindered by fences, would carry his racks of bottles across four or five backyards, delivering as he went. When he appeared between two houses to get another load of milk, there would be his horse and wagon. The horse soon learned the route and knew when to advance on his own and when not to move ahead.

A TWO-HORSE ICE WAGON. Note the rear step is folded in an up position. The bracket that extends from the rear of the body is actually a hook on which to hang an ice scale. This wagon is built to haul tonnage.

SECURING AN ICE SUPPLY. The first step was to cut the ice into blocks. Here, on the Milwaukee River, horses are seen pulling saws that cut partway through. The horse in the center of the photo is scraping snow and ice shavings off the surface.

ICE CUTTING. In this smaller ice cutting operation, large buck saws are used to cut blocks free. The blocks are then pulled out of the water with tongs and loaded on the bobsled for hauling to the ice house.

BUILT IN 1908. A one-horse ice wagon, well-painted and carefully lettered, bult by Theo. Habbegger in Milwaukee.

ICE DELIVERY. During the warm weather months ice would be delivered to businesses and homes to keep the ice boxes filled.

60

CLASSY OPERATION. Spiffy rigs driven by uniformed drivers delivered ice blocks to residential neighborhoods. Cards provided by the ice company were placed in the front window of the home to indicate whether to deliver 25-, 50- or 100-lb. blocks.

61

READY TO ROLL. These mail wagons were used in Milwaukee until around 1930 for handling certain facets of the daily routine.

RFD. The Postmaster General said in 1892, ''free delivery of mail to farmers is impossible until we have better roads.'' Four years later the Post Office Department started Rural Free Delivery. By 1912 there were 40,000 routes in the country all essentially serviced by vehicles such as this. An American flag decorates the side of this one-horse rig. Hinged doors allowed for easy entrance and exit for the mailman.

The Kelk Carriage Works of Sedalia, Missouri, said the mail wagons in their line could be kept as warm as a house in winter, were equipped with a lock box, pigeon holes, and paper holes built at an angle, table to write on, and plenty of room for bundles. They further claimed their model was worth two ordinary cheap rattle traps.

MAIL CALL. The Post Office used a variety of horse-drawn vehicles to pick up and dispatch mail. This one-horse rig is an example.

MAIL SLEIGH. Come winter the mail must go through. The runnered vehicle was a snap for the horse to pull.

WISH BOOKS ON THE WAY. This great convoy of wagons was loaded high with mail bags probably filled with colorful and exciting Sears, Roebuck mail order catalogs heading for the post office in 1906.

TRUCKING—1896 STYLE. Express wagons depart the Pabst Brewing Co. in Milwaukee with loads of promotional material and head for the railroad depot.

READY TO DELIVER. One of the trademarks of an Express wagon seemed to be the heavy wire screen sides to protect the cargo. Rigs such as this performed pickup and delivery service to commercial, as well as residential areas.

TANKER WAGON. The tank was built by Heil Co. in Milwaukee. This 4-compartment tank wagon would deliver coal oil and kerosene to the home in 5-gallon quantities.

FUEL FOR THE FARM. Farms required oil and kerosene and depended on regular deliveries.

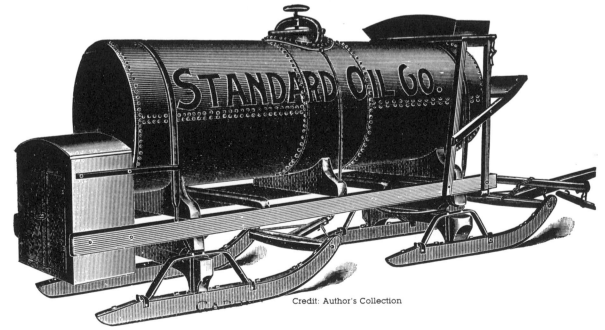

FOR WINTER DELIVERIES. Oil tanks mounted on sleigh runners were available from the Gabriel Streich Co. in Oshkosh, Wisconsin.

Credit: The Bekins Co.

THREE-HORSE MOVING VAN. This van was probably 14 feet long and had a capacity of approximately 550 cubic feet. The drop tailgate could accommodate a large quantity of bulky items.

Credit: The Bekins Co.

MOVING, 1891 STYLE. The horses stand patiently while the furniture is loaded. A wicker baby buggy is about to be carried into the van.

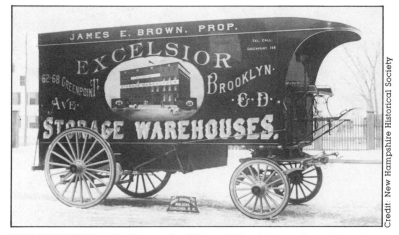

Credit: New Hampshire Historical Society

THIS FURNITURE VAN was built by Abbot-Downing Co. of Concord, New Hampshire, in the late 1880s. Many moving and storage companies had their main warehouse painted on the side of the vehicle. There was plenty of room and it was good advertising.

Credit: Author's Collection

MOVING VAN. Every city had its complement of household movers. Typical is this Kansas City operator whose letterhead is illustrated. Note the catchy slogan on the wagon: The World Moves, So Do We.

65

ROLL OUT THE BARREL. Known as a roll wagon because the beer barrels could be rolled along the tubular frame, these vehicles were built in various sizes. This was one of the largest. Note pad hanging under the wagon just behind the front wheel. The driver would place it on the street, drop the barrel off the wagon onto the cushioned pad and roll it to its destination in the saloon.

ROLL WAGON—1910 VINTAGE. Drawing of a 'roll wagon' made by Gabriel Streich Co. of Oshkosh, Wisconsin.

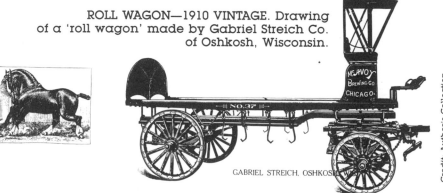

GABRIEL STREICH, OSHKOSH

MILLER "HIGH LIFE" BEER

SINCE 1855

Credit: Miller Brewing Co.

Credit: Author's Collection

HAULING CASES. These wagons were ruggedly built to withstand day-in and day-out usage. This vehicle was the product of Abresch Wagon Works, Milwaukee.

Credit: Author's Collection

LOADED WITH KEGS. The Abresch Wagon Works of Milwaukee illustrated this platform spring shipping truck in their catalog.

STAKE WAGON. Case beer delivery wagon, c. 1910.

A TEAM TO BE PROUD OF. All breweries seemed to use upstanding and good-looking horses.

BEER TO YOUR DOOR. Case beer was delivered to residential areas in smart-looking vehicles such as this.

READY TO ROLL. One day in 1907 these delivery wagons lined up in the yard of Milwaukee's Pabst Brewery for their photograph. Note the uniform good looks of these dapple grey Percherons.

WAGONS IN A ROW. The wagon shed at the Anheuser-Busch Brewery in St. Louis.

STACKED HIGH. This enormous load of wooden beer cases seems to be some sort of a promotional photo. Note the open empty case on driver's seat.

Credit: Anheuser-Busch Co.

Credit: Author's Collection

Credit: Miller Brewing Co.

BREWER'S PRIDE. Quality horses, clean harness and neat wagons seemed the trademark of all brewers.

CHOOSE YOUR STYLE. The wagon builders produced vehicles with open sides, curtained sides, or with sliding doors. These are two examples manufactured by Abresch Wagon Works of Milwaukee.

READY FOR LOADING. The really big beer wagons, when stacked high with cases of bottled beer, constituted a load requiring a six-horse hitch.

Credit: Anheuser-Busch Co.

Credit: Anheuser-Busch Co.

ANHEUSER-BUSCH. The yards of the Brewery show the stable on the right with hay loft doors open and an open air shed for wagon storage on the left. Out in the yard there is a lot of activity as 10 or 12 horse-drawn vehicles await orders.

Credit: Anheuser-Busch Co.

WAGON MAINTENANCE SHOPS. Anheuser-Busch maintained its own paint shop and wagon shop (on right). The equipment used by all breweries was always maintained and kept well painted.

HAULING KEGS. Deliveries were maintained by distributors to the small towns and villages.

DUMP COAL WAGON. A. Streich & Bro. Co. of Oshkosh, Wisconsin, built this coal wagon with a winch that would simplify unloading. Where the wagon could back up to a basement window the need to haul buckets of coal on the shoulder was eliminated.

FRESH FROM THE BUILDER. Coal wagons, painted when new, soon lost their slick neat looks because of the dusty, dirty loads they carried. This wagon was built by Theo. Habbegger Co. in Milwaukee.

CHAPTER IV

VEHICLES FOR SPECIALIZED JOBS

SOME wagons were one of a kind. Other vehicles were built in very limited quantities. Still others, while very specialized, were mass-produced by the wagon manufacturers.

There was one fact for certain—teamsters had available vehicles that could get anything moved that needed to be moved.

Portable units such as popcorn wagons or cafe wagons were very specialized. If need be, the owners took their business home with them at the end of the day.

When it came to the circus, each and every wagon was not only specialized, but one of a kind, built for hard use, and to carry an enormous or unusual load.

The hearse, police patrol and ambulance were specialized vehicles, but because every community needed them, they were produced in large quantities.

CAFE WAGON. T. H. Buckley Lunch Car Co. of Worcester, Massachusetts, developed an interesting and very specialized business. Elaborately decorated and lettered, these wagons were attractive wherever parked. C-1900.

FRESH AND HOT. Mr. Jewett of Huntington, West Virginia, parked his popcorn wagon on a popular street, then unhitched his horse and probably tied it in a vacant lot nearby. Now he is in business selling fresh roasted peanuts and fresh buttered popcorn.

THE POPCORN MAN. Dobbin stood patiently, hours on end, while Charley Hanision conducted his nickel business. Every Sunday this wagon would appear at the same spot on Milwaukee's Lake Drive. Neighborhood kids, strollers, and passing cars loaded with families would patronize Charley, who had been coming here for 38 years.

READY FOR BUSINESS. The C. Cretors and Company manufactured an entire line of specially built popcorn wagons. This 1913 Model A. weighs 2300 lbs. and cost $1,575.00 (including driver's seat) FOB Chicago. The catalog offered terms of $775.00 cash and 13 notes for $60.00 and one for $20.00. A 5% discount was allowed if the full price was paid in cash.

74

BEAUTIFUL WINDOWS. Another of Buckley's "quick lunch" wagons.

RESTAURANT ON WHEELS. Interior view of a Buckley Lunch Wagon. Ingeniously constructed, practical and efficient, they served a fast meal for those in a hurry.

PHOTO STUDIO. An enterprising photographer carried his business with him as he covered small towns and rural areas.

DISINFECTING VAN. This strange-looking vehicle with a smokestack was built in 1891 when Milwaukee's Health Department had it constructed as a ''disinfecting'' van. If a family developed a contagious disease the van was taken to the home and all furniture, carpets, bedding and clothing were placed inside. Its construction was double-walled tin with asbestos in between. Heat was the disinfecting agent. Temperatures rose to 250° and the contents were left in for 30 minutes. The heating equipment was built into the front end of the vehicle.

PULLING FOR THE DEMOCRATS. An imaginative mind came up with this billboard on wheels. The horse-drawn rig did not help the Democrats in this 1914 Milwaukee election, as the Socialist, Dan Hoan, won the mayor's office and remained in that position for 26 straight years.

STAR-SPANGLED WAGON. Political rallies and Fourth of July parades meant ordinary wagons were dolled up for the occasion. These floats were colorful and attracted attention.

HAULING TRASH. In the early 1900s, Schenectady, New York used this type of wagon—with a tilt-up dump body—for ash and rubbish collection.

IN A PICKLE. Heinz developed this eye-catching specialized rig to advertise their products.

Credit H. J. Heinz Co.

Credit: Illinois Bell Telephone Co.

PHONE SERVICE. In 1911 the Illinois Bell Telephone Co. had specialized vehicles for their servicemen.

CABLE CAR ACCIDENT. The horse pulling the ambulance gallops up to the scene. A pedestrian has been hit by a cable street car on Broadway in New York. This dramatic drawing appeared in *Leslies Weekly*, Aug. 29, 1895.

LINEMEN AT WORK. A horse-drawn overhead wire repair wagon served a special purpose. The top platform could be swiveled into position.

EMERGENCY VEHICLE. Ambulance used in New York City in 1896 by Bellevue Hospital.

PADDY WAGON. Police patrols did not look special, but they were designed to carry their passengers with security.

PRIVATE AMBULANCE. This vehicle was more attractive and elaborately decorated than those operated by hospitals.

FERDINAND 11217.

EMBALMING WAGON. This photo from 1900 shows an undertaker's embalming wagon used in Lansingburgh, New York. Note interesting fly net on ears of the horse.

A SOMBER APPEARANCE. Horses used by undertakers were generally black. The entire rig had a dignified look.

INTERIOR VIEW. Floor rollers and guides kept casket centered.

GLASS-SIDED HEARSE. Woodcarved decorations enhance the top of this plate-glass-sided hearse. Hearses used for the funerals of children were much smaller and were completely white.

ORNATE HEARSE. In 1900 this hearse cost $2,000.00. It was advertised to have "graceful carved draperies. This car makes a nice massive appearance and is constructed to run light. It is not too heavy for long drives or hilly country. Fine enough for the largest towns."

CHAPTER V

GETTING THE HEAVY JOBS DONE

It WAS ALL DONE with horse power. It might be a 70-ton steel girder to be moved to the site of a new building. Perhaps 20 tons of salt or six tons of sand to be hauled—not just today, but every day. Maybe the freighter was called to transport a 20-ton marble column for the new bank from the railroad yards to where the bank was being built.

Freighters in western mining areas were called upon to haul appalling loads—massive ore-crushers or other mining equipment or enormous boilers—up narrow, twisting mountain trails. Whatever the load, horses were called upon to get the job done—and they did!

A freighter's 10-horse team and rig would add up to a cost of $1,500.00 for horses, $750.00 for harness and $500.00 for the wagon—a cost of less than $3,000.00 for the whole outfit. Today, it would take a $100,000.00 truck to do the job—and that truck would be helpless on the roads the horses had to navigate.

All the seemingly impossible tasks illustrated in this chapter were accomplished with the brains and brawn of horses. The photographs show the loads, the wagons and the teams.

NOT AN EASY JOB. It took brawn and brains to properly load and secure an object like this. In addition, it took horsepower and skillful drivers to move the loaded wagon to a given site.

WORKING FOR THE PHONE COMPANY. In this 1911 photo a cable reel is being hauled for the A.T.&T. Co.

HAULING GOODS IN THE WEST. Frances and Dorothy Wood wrote a biography about their father, David Wood, titled, "I Hauled These Mountains In Here". David Wood was in the freighting business in Colorado in the 1880's. It was big business. In 1884 he hauled in to Ouray over two million pounds of miscellaneous freight and over one and one-half million pounds to Telluride. His waybills called for freighting such items as barrels of salt, kegs of beer, bales of wire, kegs of syrup, grindstones, cookstoves, blasting powder, tinware, steel, hardware, hay, oats, corn, flour, candles, brooms, whiskey, printing presses, coal, sugar and kegs of nails. In this photo a team with a mixed load of freight is about to take off over the rough trail called a road.

HUGE JOB. The Oregon Transfer Co. of Portland moved this gigantic boiler with 6 horses.

MASSIVE EQUIPMENT. Huge pieces of equipment were pulled to the site when the Pittsburgh gas system was being built.

IG BEAMS. When Chicago's Chas. A. Stevens Department tore was under construction around 1914, the normous, bulky and weighty steel girders were hauled hrough the city streets to the site by horse power. Twenty-wo horses moved this particular load.

A 12-HORSE LOAD. In the early part of the 20th century, the Pennoyer Transfer Co. was called upon to haul some unusual loads. Here is one of their rigs pulling an unwieldy boiler through the streets of Chicago. The twelve well-matched horses are hitched 4-4-2-2.

TRANSFORMER TRANSPORT. The weighty transformer is being hauled for the Baltimore Gas and Electric Co.

REINFORCED WAGON. What appears to be an enormous armature is being moved on this special wagon.

PERCHERON POWER. Judging from the oil derricks in the background, this 8-horse team is maneuvering an oil storage tank onto the property.

PONDEROUS LOAD. This team of 14 horses is lugging heavy parts for a dredge in Idaho City, Idaho 1910.

TIMBERS FOR THE MINE. David Wood teams transported timbers to the Virginius Mines above Ouray, Colorado in 1888. Mr. Wood said that when hauling heavy machinery on a downgrade he braked the load by snubbing a rope around a tree. The other end was fastened to the wagon. He added that frequently on downgrades they would drag logs behind the wagons to brake them.

TWENTY-TON PILLAR. In 1909 when the Bank of Sheboygan (Wisconsin) was in the process of erecting a new building, the massive marble columns arrived from a Georgia quarry at the railroad siding. The marble pillars were 26 feet long, 4 feet in diameter and weighed 20 tons each. An 8-horse team moved the pillars to the site.

MULE POWER. This view on the Erie Barge Canal shows the changing of teams on a canal boat. The general custom was for the boat to carry its mules in stables built in the bow. Here the relief team has been brought on-to the tow path to join the team that just finished its 4-hour stint. One of these mules is being led into the stable. This was steady and comparatively easy work with good footing and mules lasted 18 to 20 years on the job. Stiff head winds increased the load on the mules.

 TOUGH TRAIL. In the Western mountains, mining machinery had to be hauled up narrow roads and sometimes over trails where there were no roads. C-1900.

FULL LOAD. In Miss Frances Wood's book, David Wood explained: "A forwarding house was one that received freight from the railroad, paid the freight thereon, and then shipped the goods by freight wagon to people in the mountains or valley towns. On the return trip, the teams usually brought out ore to go to the smelters. We had two wagons in tandem and figured six mules for 5 or 6 ton; eight mules for 8 or 9 ton, and 10 mules for 10 or 11 ton of ore. In 1884 we hauled over 3 million tons of ore out of Ouray and Telluride to Montrose." This photo shows a typical David Wood outfit.

HARD AT WORK. Construction of a building started with the excavation of basement and foundation area. Here, in 1900 at the northwest corner of Milwaukee Street and East Wisconsin Ave. in Milwaukee there is much activity taking place. The team in the center of the photo is standing on the bridge over the wagon. Its loaded road scraper will be dumped through an opening into the wagon below. When the wagon is full it will be moved down to the wall where a donkey engine will haul it up to street level. Note team of white horses on the street waiting for a loaded wagon. The black team in the pit is fastened to a huge hook-like plow that loosens the dirt so the team with the road scraper can easily scoop it up.

TWO KINDS OF POWER. A Bucyrus steam shovel scoops up the dirt, swings around, and dumps its bucket load into the wagon. When loaded, this wagon with its dump bottom will haul the dirt to the assigned spot. Another team is just arriving and will back into place to receive another load.

KEEP IT MOVING. The toll road between Ouray and Silverton in Western Colorado in 1888 became blocked by a snow slide in early spring. This did not stop the freighters for long. They tunnelled through the snowmass and kept traffic moving.

Credit: Frances Wood Collection

LOGGING SCENE IN NORTHERN WISCONSIN. A story in the Dec. 27, 1911, issue of the ''Breeder's Gazette'' said, ''Horses in logging camps are usually geldings 6 to 12 years old. Skidders—those horses that are used to haul logs from place where the tree has been felled to the skidway should weigh 1400 pounds to 1600 pounds. Many of these horses last 15 years on the job. The drivers groom both night and morning. The barn man feeds them and mends harness. The horses are fed 15 pounds of timothy a day, and 6 quarts of oats three times a day. Every Sunday the horses got a hot bran mash. The horses are shod with sharp toe calks and small heel calks.''

UPHILL JOURNEY. The Kramer Wagon Co. of Oil City, Pennsylvania, had this page in their catalog.

NEW YORK, 1912. Here three 3-horse teams of Percherons are lined up to receive massive loads of rock.

LOG SLED.
The sleigh runners are wide apart to accommodate the load and keep it from being top heavy. Frequently, the roads were watered down to ice them, thus easing the pull for the horses.

A LONG LOAD. This extra long and huge oak timber is being hauled out of the woods in Monroe County, Ohio. A team like this, one black and one white, was frequently referred to as a Boston Matched Team.

HAULING BOARDS. Freshly sawed lumber on the way from the sawmill to the railroad cars in northern Wisconsin.

"GREAT MONSTER OF A LOAD." These two pages, in the catalog (C1910) of the Gabriel Streich Co. of Oshkosh, Wisconsin, tell a fascinating story.

GABRIEL STREICH, OSHKOSH, WIS.

No. 117.   Boss Logging Sleigh.   Latest improved Logging Sleigh, built in three sizes, with 3½x8x8 feet,
White Oak Runners, Single Beam.   3½x9x9 feet, White Oak Runners, Single Beam.   4x9x9
feet, White Oak Runners.   Width of Track, 6 feet to 7 feet, center to center of Runners
with swinging Bunks, front and rear.   State dimensions, width of track from
center to center of Runners.   Write for prices.

106

TESTIMONIAL.

A Monster Load of Hemlock.

From the John R. Davis Lumber Company's Camp
No. 2. Scale 16,860 feet: Containing 114 Logs.

All of Phillips turned out Saturday to see the big load of logs which was brought in from camp No. 2 of John R. Davis Lumber Company, of which Joe Hunter is foreman.   The load was expected in on Saturday afternoon, but did not arrive here until late in the evening.   The delay was caused by having to clear out the trees which stood to near the road and getting up one large hill, where we understand, it required thirteen teams to pull it.

This great monster of a load was loaded by Joe Hunter and John Murphy and consisted of 114 logs, which scaled 16,860 feet.   It was a marvelous piece of work and one wonders how they got them there as the load was all of 20 feet high.   This gigantic load of Hemlock was hauled a distance of twelve miles by four horses, which were driven by A. LaFountain, except up one or two hills where more horses were required.   On one of these hills it required twenty-six horses, so we are told.   The sleighs were made by Gabriel Streich of Oshkosh, and carried a twelve-foot bunk so that the base of the load was about fifteen foot wide.   While this is not the largest load ever hauled in this vicinity, yet it is the largest ever hauled so great a distance and is certainly the largest for this winter.

PUBLISHED IN "THE BEE," MARCH 21, AT PHILLIPS, WIS.

107

CHAPTER VI

PUBLIC TRANSPORTATION

LONG BEFORE there were railroads and Greyhound and Trailways bus lines, there were stage coach lines carrying people and their baggage from one end of the United States to the other.

Long before there were city buses and trolley cars, there were omnibuses and horse-drawn street cars.

Long before there were taxis, there were horse-drawn hacks.

Long before Hertz and Avis, there were livery stables.

Long before there were station wagons, there were tallyhos and wagonettes.

All things considered people got around in good style from neighborhood to neighborhood, from town to town, state to state, and even ocean to ocean by horsepower.

Stagecoaches Kept the West Rolling

Back in the 1860s Ben Holladay operated the Overland Stage Co., a system that covered 5,000 miles from Kansas to California with branch lines to the gold camps and various mines in Colorado, Montana, Utah, Idaho, and Oregon.

Overland Stage Co. operated about 500 coaches and approximately 500 freighting wagons. All this equipment required 5,000 horses and mules, 150 coach drivers, plus 300 hostlers and stock tenders.

On Overland's cross country run they had

Credit: Author's Collection

HANSON CAB. This vehicle would carry two people. Note the levers which opened or closed the doors. C-1895

MADISON SQUARE PARK, 1901. The drivers of the hansom cabs were patiently waiting for fares.

PUBLIC FOUNTAIN. Cabbies drive their horses to drink. Facilities like this were found in every city or town, big or small.

PUBLIC TRANSPORTATION

50 stations between Atchison and Denver; 51 between Denver and Salt Lake; and 55 stations from there to Placerville, California.

In 1863 the ride from Atchison to Denver cost $100. It took six days and six nights to travel the 652 miles with 50 stops to change teams.

Under Holladay, president of the line, there was a General Superintendent of the entire line, a Division Superintendent in charge of every 600 miles and a Division Agent in charge of each 200-mile segment.

The Division Agent was always on the go as he checked out the stage schedules, the stations and their keepers and meals served patrons and stock. He had to purchase grain and hay and hire employees—drivers, stock tenders, station agents, harnessmakers, carpenters and blacksmiths. Most important, he had to be able to get on the coach and drive the six-horse team in case of an emergency.

Swing stations were located about 10 to 15 miles apart. Depending on the terrain, the horses were kept at a trot or gallop practically the entire run, with stops only long enough to change at each station.

Home stations were 25 to 35 miles apart. There, the drivers were changed. At meal stations, the coach would stop for 20 minutes so passengers could eat. Schedules were kept to the point the agent's wife could have the meal ready when the coach rolled up in a swirl of dust.

The blacksmith moved up and down the line constantly shoeing the horses or mules and checking their feet.

Generally, the coaches were Concords built by Abbot-Downing Co. of Concord, New Hampshire. Painted a bright orange-red with yellow wheels and undercarriage, beautifully striped, these vehicles flashed through

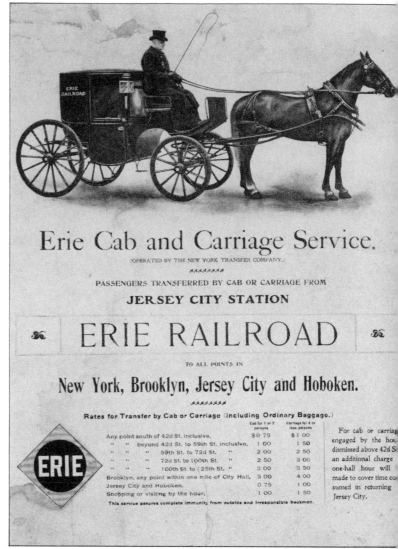

CAB AND CARRIAGE SERVICE. The Erie Railroad had its own cab service at the Jersey City Station.

CHICAGO, 1888. Cabs jam the street outside Chicago's Auditorium where the Republican National Convention was being held. They nominated Benjamin Harrison.

STAGE LINE.

ROCK ISLAND
to
CHICAGO.

IN FOUR HORSE POST COACHES,
and
BY RAIL ROAD.

Leaves Rock Island daily, at 4 o'clock, A. M., via Moline, Hampton, Port Byron, E rie, Lyndon, Como and Sterling to *Dix-on* and *Chicago, in 36 hours.*

Passengers who wish to take the Mail Stage, will land at Port Byron, as the River can not be crossed at Meredosia, with safety, and the Roads and Bridges being impassable, it compels us to omit Albany with Chicago Stages, for the present.

The above Line connects with all the Stages out of *Dixon* to

PERU, ROCKFORD and GALENA,

And it also connects with the Cars at Aurora, daily.

Chicago, Aug. 20, '51. **JOHN FRINK & Co.**

☞Also, daily Lines of Stages from Davenport, Iowa, to Dubuque, Muscatine and Iowa City.

JAS. J. LANGDON, PRINTER, 161 LAKE ST., (2d & 3d Stories,) CHICAGO.

1851 poster advertising the Stage Line from Rock Island to Chicago, a 36-hour trip.

PUBLIC TRANSPORTATION

MONTANA STAGE COACH. The stage in Wolf Creek Canyon shows signs of hard use. Lady passenger proudly displays her baby for the photographer.

the countryside drawn by four or six horses, depending on the grades. The teams were generally matched—all bays, greys, or perhaps browns.

The Overland Stage Line, in the 1860s, allowed each passenger 30 lbs. of baggage free, but charged $1.00 a pound for each pound over that amount.

A regular Concord coach had seats for nine people—three facing forward, three facing backward, and three on jump seats in the middle. Some Concords had seats for 12. Some had seats on top.

There were many other lines that operated in various areas, including Butterfield & Co. which ran from St. Louis to El Paso and the Wells-Fargo Express Co.

Horse-Drawn City Transport

In the cities, the horse-drawn omnibus would carry people through the main streets. Sometimes these were double-decked vehicles drawn by three horses abreast.

Then came horse-drawn street cars that moved on rails. More people could be handled on one car, but it was difficult for a team of horses with steel shoes to get an overloaded car started, particularly on brick pavement.

Horse cars could move people at an average of about five to six miles per hour unless the car got hung up in congested streets.

In New York City, in 1890, the Street Railway service was using 15,000 horses. These animals lasted perhaps three to five years before they went lame from the constant pounding on brick streets.

(Settlers in Kansas and Nebraska could purchase worn out horses from the Chicago Street Car Company at low prices. They turned them loose to run on the soft earth, and the horses soon regained their usefulness and became wagon and plow horses for the settlers.)

The horse cars were equipped with hand-

"Adherence to the following rules will insure a pleasant trip for all.

"1. Abstinence from liquor is requested, but if you must drink, share the bottle. To do otherwise makes you appear selfish and un-neighborly.

"2. If ladies are present, gentlemen are urged to forego smoking cigars and pipes as the odor of same is repugnant to the Gentle Sex. Chewing tobacco is permitted, but spit with the wind, not against it.

"3. Gentlemen must refrain from the use of rough language in the presence of ladies and children.

"4. Buffalo robes are provided for your comfort during cold weather. Hogging robes will not be tolerated and the offender will be made to ride with the driver.

"5. Don't snore loudly while sleeping or use your fellow passengers' shoulder for a pillow; he or she may not understand and friction may result.

"6. Firearms may be kept on your person for use in emergenies. Do not fire them for pleasure or shoot at wild animals as the sound riles the horses.

"7. In the event of runaway horses, remain calm. Leaping from the coach in panic will leave you injured, at the mercy of the elements, hostile Indians and hungry coyotes.

"8. Gents guilty of unchivalrous behavior toward lady passengers will be put off the stage. It's a long walk back. A word to the wise is sufficient."

PROPER BEHAVIOR. Wells-Fargo Stage Line issued these rules for passengers riding in their coaches.

FULLY EQUIPPED. In 1879 the Fish Creek stage station in Madison County, Montana, looked rather forlorn. It did have all of the necessary ingredients. The building on the left was a place to eat and also a post office; in the center is the home of the station manager; on the right is a feed barn. Corrals for horses are also seen.

READY TO ROLL. A stagecoach about to pull away from the Wells-Fargo office in Helena, Montana in 1866. The rack on the back of the coach, called the "rear boot", carries luggage and express. It is seen here covered over with a leather apron to protect it from the elements, including the rolling dust in dry weather.

PUBLIC TRANSPORTATION

brakes to hold the cars back on downgrades and also to stop the car at assigned stops.

Most major cities had horse-drawn street cars. In 1875 Cincinnati had 45 miles of track on which 160 cars rolled. The stables held over 1,000 horses.

At the end of the line the driver detached the pole, brought the team around to the other end of the car, inserted the pole in its socket, dropped in the pin, and was all set for a return trip.

In May of 1887, New York City had a terrible tragedy when one of their street car stables was destroyed by fire and 1,600 horses perished.

By 1890, when electric street cars began to be seen on city streets, the horse cars were on the way out.

Many cities had sightseeing services. Groups newly arrived in town by excursion boat or train would get on a huge tally-ho drawn by six horses. A ''rubber-neck wagon'' could carry 20 to 30 people.

Rigs for Rent

Livery stables—and there were many in every city, quite a few in every town and usually one in every hamlet—had two classes of patrons: people who hired a horse and buggy for various reasons, and people who boarded their horses. This latter group had no stable facilities at their homes, or possibly could not afford stable upkeep, hired man, feed, vet care, etc.

In Milwaukee, the city code required an annual license fee from the liveries of $50 for a tally-ho drawn by four or six horses. The code also had a $5.00 annual fee for a carriage and team.

Any carriages, said the code, that were waiting at curb at night for a fare had to have a lighted lamp attached.

The city also set the cab fares including

FILLED TO CAPACITY. An Abbot-Downing Coach in the 1880s carrying passengers from Glens Falls, New York to Lake George. This oversized vehicle handled quite a load—17 people can be seen on the roof.

SAN JUAN STAGE, MAIL AND EXPRESS LINES.

TIME CARD.

No. 1.   IN EFFECT JULY 1st. 1887.

FOR USE OF EMPLOYEES ONLY

MONTROSE TO OURAY.

		Distance
Montrose, Leave 7 a. m.		
Fort Crawford,	Arrive 8 15 a. m.	9 Miles
Los Pinos Agency.	" 9:30 "	13 "
Cow Creek,	" 10:45 "	20 "
Dallas, - Dinner,	" 11:30 " Depart, 12:00 m.	25 "
Ramona,	Arrive, 1 15 p. m.	33 "
Portland,	" 1:25 p. m.	34 "
OURAY,	" 2:00 p m	38 "

MONTROSE TO TELLURIDE.

		Distance.
Montrose, Leave 7 a. m.		
Wells,	Arrive 9:30 a. m.	12 Miles
Johnson's,	" 11:00 "	20 "
Horsefly,	" 11 15 "	22 "
Smith's, Dinner	" 12:30 p. m.	30 "
Placerville,	" 2:30 "	37 "
Wheeler's,	" 3:15 "	41 "
Sargent's,	" 4:45 "	49 "
TELLURIDE,	6:00 "	55 "

OURAY TO MONTROSE.

		Distance.
Ouray, Leave 1 p. m.		
Portland,	Arrive 1:30 p. m.	4 Miles
Ramona,	" 1 40 "	5 "
Dallas,	" 3:00 "	13 "
Cow Creek,	" 3.45 "	18 "
Los Pinos Agency,	" 5 00 "	25 "
Fort Crawford,	" 5:30 "	29 "
MONTROSE,	" 7:00 "	38 "

TELLURIDE TO MONTROSE.

		Distance.
Telluride, Leave 7 a. m.		
Sargent's,	Arrive 8 00 a. m.	6 Miles
Wheeler's,	" 9:30 "	14 "
Placerville,	" 10:00 "	18 "
Smith's, Dinner),	" 12 00 m	25 "
Horsefly,	" 2:00 p. m	33 "
Johnson's,	" 2:30 "	35 "
Wells,	" 4:00 "	43 "
MONTROSE,	6:00 "	55 "

GUNNISON AND LAKE CITY ROUTE.

Gunnison, Leave 7 a. m.

		Distance.
Powderhorn,	Arrive 12 00 m. Depart, 12:30 p m.	26 Miles
Allen,	Arrive, 2:30 "	35 "
LAKE CITY,	" 7:00 "	56 "

Lake City, Leave 7 a. m.

		Distance.
Allen,	Arrive, 11:30 a. m. Depart, 12:00 m.	21 Miles
Powderhorn,	Arrive, 2:00 p. m.	30 "
GUNNISON,	" 7:00 "	56 "

CONNECTIONS.

At OURAY close connection is made with Ouray and Silverton Stage Line for Ironton, Red Mountain, and Silverton.

At TELLURIDE with Western Mail & Express Company's Stages for Ames, Ophir, Trout Lakes and Rico.

At LAKE CITY with coaches for Capital City, and Rose's Cabin.

Ask for THROUGH TICKETS at Denver & Rio Grande Railway Ticket Offices.

DAVID WOOD,
PROPRIETOR.

THE COLORADO CONNECTION. David Wood, who also hauled freight, ran a stage, mail and express line. This is one of his time cards giving times of departure and arrivals, as well as distances.

DAILY
OVERLAND MAIL

☞ THROUGH IN TWENTY DAYS! ☜

SACRAMENTO, Cal., to ST. JOSEPH, Mo.
— VIA —
PLACERVILLE,
CARSON CITY, and
SALT LAKE CITY.

TARIFF OF FARES:
FROM SACRAMENTO

TO CARSON CITY $25	TO FORT LARAMIE.........$155
" FORT CHURCHILL....... 35	" JUELSBURG,—CROSSING
" SAND SPRINGS........... 40	OF SOUTH PLATTE .. 155
" FORT CRITTENDEN.....100	" FORT KEARNEY......... 155
" SALT LAKE CITY........110	" OMAHA.... 155
" FORT BRIDGER...........125	" ST. JOSEPH 155

The Mails and Passengers will lay over one night at Salt Lake City.

☞ Passengers will be permitted to lay over at any point on the road, and resume their seat when there is one vacant. To secure this privilege, they must register their names with the Stage Agent at the place they lay over. Passengers allowed 25 pounds of Baggage; all over that weight will be charged extra.

The Company will not be responsible for loss of Baggage exceeding in value Twenty-Five Dollars.

FOR PASSAGE, APPLY AT THE

STAGE OFFICE, Second Street, between J and K,
— TO —

H MONTFORT, Agent,
SACRAMENTO.

TWENTY-DAY TRIP. The Overland Stage would take a person from Sacramento, California, to St. Joseph, Missouri, in 20 days at a cost of $155.00.

DRIVE CAREFULLY! The toll road in the Red Mountain area of western Colorado was certainly scenic and probably hazardous.

The First Americans to Rock 'n' Roll

They Were Hardy Pioneers Who Endured Days of Extreme Discomfort as Passengers in a Bouncing, Rattling 'Piece of Joinery' Called the Stagecoach

FIRST you saw a burst of dust far off where the turnpike entered the woods. Down the wind came the pop of the stage driver's whip, and his bawling "HeeeYawww!" as he urged his teams for a flourishing arrival.

As the big Concord stage rocked closer, you could see sweat on the straining horses and the sparkle of brass and varnished wood. There was a jangle of harness and a long screeching of brakeblocks. Then, wondrously, the stagecoach had arrived, smelling of cigars and hot gravel and trailing the magic of distant cities.

By this time the street was full of barking dogs and running children. Even the old timers stopped whittling to greet the driver as he swung down from his high seat, beating dust from his hat. Sometimes the passengers didn't even have time to join him for a quick belt of "Old Monongahela." The boast along the National pike was that a good groom could change teams before a coach quit rocking. Then, if it was an express run, it was "Hi-yup!" and away again with snapping whip and a scattering of small boys and chickens.

Today, with jet planes and superhighways, there is no counterpart for the once popular pastime of "watching the stage come in." Nevertheless, during the heyday of the stagecoach—from about 1830 to 1900—there was no greater thrill for an isolated town than to welcome a stage bringing mail, new faces, fresh news and, most of all, a visible tie with the outside world.

British Invented Stagecoach Name

Nobody knows who invented the stagecoach. Persian wall paintings dating from 430 B.C. show a vehicle called the harmanaxa, which combined the idea of four wheels and a box body. The French made coaches similar to stagecoaches as early as the 15th century. The English began making such coaches a century later, and invented the name "stagecoach" to identify a commercial coach that moved by stages. Scotland had a stagecoach line as early as 1610, and England was laced with them by 1700.

Coaches were made in the American colonies at least as early as 1687. By 1755 stagecoaches linked Boston, New York and Philadelphia, and provided service that amazed their customers.

"We were rattled from Providence to Boston (40 miles) in 4 hours and 50 minutes," the editor of the Providence Gazette wrote with pride. "If anyone wants to go faster he may send to Kentucky and charter a streak of lightning."

Stage lines extended south to Georgia by 1803, and a year later had thrust west to Pittsburgh. An express firm dubbed the "shake gut" line boasted that it could bring Maryland oysters to Pittsburgh so fast that its customers "had no fear of being poisoned." The famous Cumberland, or National pike, a good gravel road, was started in Maryland in 1806, and by 1838 had reached Vandalia, Ill. A few years later Wisconsin had stage lines between Milwaukee, Prairie du Chien, Eau Claire and Green Bay and north to Houghton in upper Michigan.

All manner of horse drawn rolling stock was tried on these roads. There was the "mud wagon," the "celerity wagon," the "tally-ho," the "barge," the "drag" and the "jerkey." Some of these survived—the celerity wagon was fast, and the light, high wheeled mud wagon could bounce through anything. But the best known commercial transportation vehicle of its day was the Concord coach, built in Concord, Mass., by the Abbot-Downing company.

The Concord was the kind of stagecoach you see on television—a big high wheeled gullumpher, with an egg shaped body, luggage "boots" fore and aft, a driver on top and six horses pulling. It was, as one historian says, "quite a piece of joinery," with its seasoned ash body, white oak underpinnings and hickory spoke wheels.

Concord Coach Carried Eleven

The standard Abbot-Downing color scheme was a barn red body, varnished to gleaming, and yellow running gear. Coaches carrying mail were emblazoned with a big "U.S. Mail" in gold letters. Some had fancy flower and twining vine designs; many bore painted pictures of famous actresses below the driver's seat. There was a pair of cut glass coach lamps, mostly for show, and on top a brass railing for lashing down wooden trunks and other bulky freight. In the west there was often an iron cash box, riveted to the floor in the boot below the driver.

Official capacity for a Concord was 11 paying passengers, with nine inside on thick cowhide spring cushions, and one on each side of the driver. Usually people simply crammed aboard until there was hardly room to nurse a baby. Loads of 15 or more were common, with the short run record said to be 35.

Besides cushions, the passengers enjoyed such comforts as safety belts and grab loops for rough going, roll down leather curtains, basins for coach sickness, glass windows (sometimes) and iron footstoves and hot soapstones in winter. There was also the Concord's famous leather springing, which kept the coach fairly level and was alleged to transform bumps into "a velvety sway." This unique suspension consisted of two long straps of eight ply sole leather which supported the coach as in a hammock. There was no give in the supporting stanchions, however, and a deep chuckhole could knock your hat off.

The first Concord was built about 1827. As late as 1900, one writer called it "the only perfect passenger vehicle for traveling that has ever been built." They cost from $1,200 to $1,500 (there were heavy and light models) and were exported to South America, South Africa, Australia and even mother England.

Stage Driving Took a Knack

Like a ship captain, the driver, or "jehu" ("for he driveth furiously"—Kings 9:20) ruled supreme. Drivers tended to be red nosed, tobacco chomping types, with names like Baldy Green, Old Charlie, Jared Crandall and Hank Monk. The calling became hereditary along the National pike, as did the profanity and tavern stories.

Stage driving was prestigious work. As one bleary eyed jehu informed an English passenger, "When I drive this stage I am the whole United States of America." It took a knack to manipulate six reins in one hand while cracking a whip with the other and stamping the right foot on the long brake lever. The drivers also appreciated the value of spectacular packaging, lavishing their often ratty horseflesh with fake tails and brass and ivory gewgaws. To be invited to sit next to the driver was an honor given to senators, pretty girls and guards "riding shotgun."

The most exciting era in stagecoaching began in 1858,

when John Butterfield's Overland Mail Co. offered the first dependable service between St. Louis and the west coast. Bound to a government contract, the Butterfield stages covered the distance in an average 23 days, stopping only to change horses, drivers and passengers. A passenger, who paid a $200 fare, not including meals, could stay aboard as long as he could stand it. Several people went insane trying to tough out the whole stint before drivers began forcing people to rest over occasionally and catch a later coach.

The worst part of such travel was sleeplessness, especially in a crowded coach. Other hazards included smothering heat, burning alkali dust, corduroy roads, drumming hailstones, collapsing bridges, rushing fords and whooping Indians. On a steep hill, passengers often had to hop out and hold drag ropes while driver and stage skidded down on chain locked wheels. When the stage got stuck in a mud hole, they had to help push and pry with heavy fence rails. Another duty was to lean right or left on shouted command ("Now, gentlemen, to the left!") to keep the coach from tipping. Little wonder that Horace ("Go west, young man!") Greeley once staggered out of a stage too exhausted to lecture on his favorite subject.

'Home' Stations
Had Meals, Bunks

The 141 stations on Butterfield's line were divided into "way" stations and "home" stations. A way station was usually a ranch house or sod dugout with little more than a muddy well and a couple of cottonwood trees for amenities. The home stations provided meals, nickel whisky and rope slung bunks with ticks full of "prairie feathers" (dried grass).

After a day rubbing knees with toothless prospectors and unwashed gamblers, a lady might look forward to a supper of dried beef and boiled hominy, to be followed next morning with biscuits "hard as roundshot" and tamales "which appeared to have been ground from blue flint and sawdust." The legs of the beds commonly stood in pots of water to keep the bedbugs from spreading, but it was a losing fight.
"They shore do be bad," a station hostess whinnied to a guest, after squashing one with her bare toe.

Coaches always seemed to arrive before dawn, forcing the traveler to get up in the dark and stumble around with a candle or horn lamp. To wash up, you used the horse trough or, if you were extra fastidious, a pan of cold water and a bar of gritty lye soap.

There were few mourners when a spreading railroad system pushed the stagecoach off the road and into the museum. The last of them survived until about 1910, hauling mail and a few passsengers into remote towns in the west. —*JAY SCRIBA*. *THE MILWAUKEE JOURNAL*

STAGECOACH STORY. This piece appeared in the Milwaukee Journal on April 20, 1964.

Credit: State Historical Society of Wisconsin

NICE DAY FOR RIDING. On pleasant days passengers preferred sitting on roof seats (if the coach was so equipped) to sitting in the stuffy interiors.

Credit: Ron Ryder Collection

DOUBLE-DECKER COACH. Some Concord coaches had seats on top, such as this one ready to leave Prospect House, probably in Utica, New York C-1900. These vehicles were called "Concords" because they were built by Abbot-Downing Co. in Concord, New Hampshire.

PUBLIC TRANSPORTATION

a 50¢ rate from one hotel to another. Hauling a trunk was 50¢, and 25¢ for each additional trunk.

George J. Reilly, whose father owned a livery stable in Providence, Rhode Island, wrote a fascinating story of his recollections of this kind of operation for the *Carriage Journal*. Some excerpts: "The liveries boarded many horses but they would not take on a new customer unless the former livery man gave the customer a good reference.

"The downtown liveries had rigs of all kinds, for drummers, carriages for funerals, railroad depot work, going to balls, or taking women shopping.

"One livery had a contract to furnish horses for the Post Office, a couple of newspapers and garbage pick up. He had 1,800 horses in his stables. Most stables had men on the night shift, in case horses and drivers were needed.

"The night men washed vehicles, cleaned harness and greased wheels. Depending on the work load, horses were fed, watered and groomed by both day and night shift men. Each horse was fitted with bridle and harness, then numbered according to stall number. Thus, the same horse used the same stall each day.

"Livery men wanted horses they rented out to be a little on the lazy side, short barreled and true going. Most every downtown liveryman had a contract with a big hotel to furnish rigs. The price was $2.00 a day for a

SEEING THE TOWN. To get to the roof seats on this rig, a person had to climb the step plates on the side of the vehicle. Those inside entered through a rear door.

DOUBLE DECKER BUS. Horsedrawn omnibuses carried people up and down the streets of all the big cities. Here is a double decker operating in New York.

Credit: Author's Collection

106

GETTING THROUGH THE SNOW. Come winter, omnibus
sleighs replaced the wheeled vehicles.

TURN-OF-THE-CENTURY TRANSPORT. A three-
horse double decker bus on
New York's Fifth Avenue in the 1890's.

PUBLIC TRANSPORTATION

single horse and vehicle. A single horse and surrey for a Sunday drive was $4.00 a day. If a team was required, the fee was $6.00. For funerals, hacks were $4.00 per funeral and $8.00 for the hearse. If grocery stores, laundries, etc. had their own delivery wagons, the rent for a horse was $1.75 a day.

"Mr. Reilly recalls that 'my father gave every horse in his stable a two-week layoff at some period during the summer.'

"Reilly's Livery had two horse-drawn ambulances available. We also had a service each March and November of clipping horses. The charge was $2.00 a horse. Many liveries, particularly where the owners were friends, would get together and add up their requirements for hay and oats, then purchase carload lots in order to get the best price possible."

Livery stables were central gathering points. Hangers-on sat on benches and rickety chairs out front on nice days gossiping about horses and people.

Farmers, or folks from a neighboring town, could park their horses at the livery while they conducted their business. The stable operator would feed and water the horses for perhaps a quarter.

Renting horses to people with rough hands, short tempers, and little knowledge of driving caused livery owners real headaches. Unnecessary abuse was telling on the horses.

The degree of cleanliness around a livery stable depended entirely on the manager. His problems were two-fold—manure and the dust from hay, straw, and grooming.

Harness and vehicles had to be kept clean. The horses had to be curried and brushed. The stalls had to be mucked out. The degree of cleanliness determined the degree of order. This atmosphere was a trademark of any horse stable.

BRAND NEW. A neat rear-entry omnibus, probably of the 1870s. Note the racks on top for luggage and packages. The photograph was taken in front of John Stephenson Co. in New York, the manufacturer.

EIGHT-PASSENGER WAGONETTE. The Studebaker Corp. of South Bend, Indiana, promoted this wagonette in their 1912 catalog. They said that the glass windows could be lowered.

ANOTHER 1912 STUDEBAKER VEHICLE. The Mountain wagon had rolldown side curtains in case of inclement weather, and a luggage and package rack on the rear.

EXTRA LONG. "Pride of the Nation" is beautifully lettered across the side of the body of this unusually lengthy omnibus. It was 36 feet long and pulled by ten horses. It was built in 1875 by the John Stephenson Co. in New York.

SCHOOL BUSES. Horse-drawn buses were used in 1915 in Frederic, Wisconsin.

TOURING MILWAUKEE. The sightseeing bus was provided for passengers up from Chicago on the ship S.S. Virginia. The steamship line had their own coach service. The six blacks carry their heavy load along at a brisk trot through a Milwaukee Park on this excursion in the 1890s.

FOR HIRE. Livery stables were commonplace in all towns and cities. In Frankfort, Kentucky this livery
was located on Main Street near the downtown business district. Note the buggies in readiness for itinerant customers.
The horses could be stabled on the second floor leaving the first floor for easy storage of various vehicles.
Both arched doorways were equipped with solid doors. Note the slatted areas above each door which allowed for
air circulation when the solid doors were closed either at night or during severe weather.
The windows on the upper floors were opened for additional ventilation.

SHOWING OFF THE STOCK. When it came time to be photographed, Mr. Hindman brought out some of his best horses. Livery stable operators were proud of their animals. Note the usual bench against the front of the building where men could sit and smoke and brag and boast about horses.

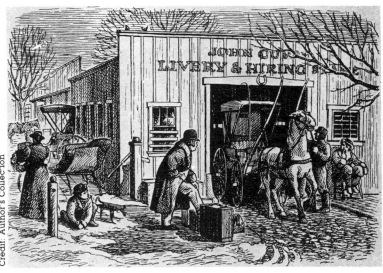

RENT-A-HORSE. The salesman with his sample cases awaits the rig he arranged to rent while the livery stable operator is hitching the horse. The livery stable owner always had a good selection of quiet, easygoing mares to supply because he never knew the quality of the horsemanship of his customer.

FULL SERVICE LIVERY. The Milwaukee livery stable of Best & Gerber on Chestnut Street in the 1890s. A stable like this would board horses for neighborhood residents, rent horses and buggies to drummers, provide extra rigs for funerals and cabs for hotels.

FIRST CLASS RIGS. The Burton House was apparently the local hotel and operated their own livery. The worn and faded sign on the side of the shed says, ''First Class Rigs and Saddle Horses— Open day and night.''

111

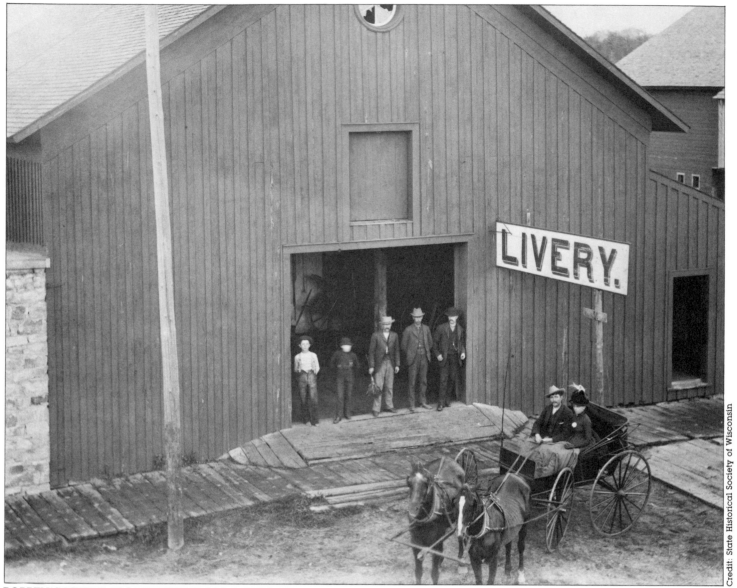

POPPEN'S LIVERY, BLACK RIVER FALLS, WISCONSIN. An interesting feature is the ramp built to make for easy entrance and exit of a carriage or buggy.

ROOM FOR RIGS. The ramp seen on the right was used to take lightweight buggies and carts to the second floor for storage.

CLIFTON 20. (4689) 272.

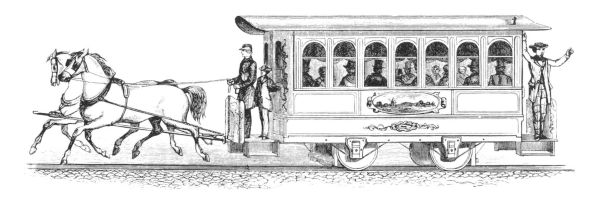

SLED HORSE CAR. If the streets were not plowed—and frequently they were not—the Winnipeg, Canada, street railway was ready to operate in the winter with their horse cars on runners.

OPEN HORSE CAR. The North Chicago City Railroad was one of the largest operators of horse cars in Chicago.

TWO LEVELS OF TRANSPORTATION. Street cars are seen here moving through the Bowery in New York in 1888. The horses quickly became accustomed to the rattle and clanking of the elevated trains.

Credit: New York Public Library

STREET CAR STORAGE. The Second Street Stable and Car House of the Milwaukee City Railway Co. in the 1870s.

Credit: Milwaukee Journal

MILWAUKEE IN THE 1880s. This city first used horse cars in 1860 and they saw service for nearly 30 years before they were replaced with electric trolleys.

HEADING DOWN. Many cities had severe hills to navigate, which required wheel brakes operated by the driver. This scene is in Cincinnati in the 1880s.

MOVING BILLBOARDS. Advertising signs cropped up early on as can be seen in this 1870 photo in Cincinnati. When the streets were not paved and became muddy it did not make it any easier on the horses since the cars did not roll smoothly.

PRIVATE TRANSPORTATION

HANDY WAGON. Whatever its name—driving wagon or runabout—it was a useful vehicle, with room under and behind the seat to carry parcels and packages.

THE CONVENIENCE of owning a horse and buggy was limited to those persons wealthy enough to afford it: merchants, bankers, lawyers and doctors.

Small towns had no public transportation systems. To get anywhere, if you did not own a horse to ride or a rig to ride in, you walked. (The bicycle changed all this. Bikes did not need stables, hay and oats, although they had their limitations during inclement weather. This form of transportation became most prevalent in the big cities because more streets were paved. Bicycles gave the working man an independence he never had before.)

In Francis Underhill's book ''Driving for Pleasure'', published in 1896 he comments, ''A good carriage is intended for many years of hard use and not to be thrown aside like a woman's gown in obedience to the dictates of any and every whim of so-called fashion. The same may be said of a well kept harness; and all this goes to show how important it is that the first choice be carefully made.''

Quality in vehicles (and harness) made a big difference, but good care was most important. There were scores of models and types and styles and many variations, some only slight, to choose from.

Most popular was the buggy—a light four-

DONKEY CART. Donkeys, while not elegant looking, supplied satisfactory horsepower.

MAKING HOUSE CALLS. Frequently, a small town doctor was called to outlying farms or villages. Doctors generally had good, upstanding horses, as seen in this photo taken near Lansingburgh, New York. The tasseled fly net bonnets on the ears could be described as "cute," but they served a practical purpose.

117

PRIVATE TRANSPORTATION

NAME THAT VEHICLE. Different manufacturers had different names for seemingly the same general vehicle. The above buggy might be referred to as a "Road Wagon" or a "Spindle Seat Driving Wagon", or, perhaps, a "Runabout". This vehicle the young ladies are enjoying near Waitsville, Vermont, has wire spoked rubber tire bike wheels.

wheeled vehicle with a top. Then there were broughams, a closed carriage for two driven by a coachman. Cabriolets were driven by the owner, and were open carriages seating two. A chaise was a two-wheeled light open vehicle for two driven by the owner. Coach is a generic term used to describe a closed vehicle that could seat four or more people. The landau was an open carriage, coachman driven. Some models were convertibles, as the top and window sides could be put in place quickly. The sport drove a sulky, a two-wheeled single seat cart. An elegant Victoria was a low-built open carriage without doors, seating two persons, coachmen driven. The vis-a-vis was a coachman-driven fancy carriage in which the passengers faced each other.

These and scores of other types gave prospective buyers an enormous variety to choose from.

Some models of the big coaches had optional items, such as speaking apparatus tube", so the passenger could communicate with the driver. Another option was a built-in toilet.

In Tom Ryder's complete book *On The Box Seat* (Pub. 1969) he points out, "There are three aids to driving—reins, whip, and voice.

"For carriages and coaches, harness may be grouped in four categories—State, Dress, Sporting and Utility.

"The State harness is elaborately embellished. Dress harness is proper with carriages driven by liveried coachmen or owner in town, or park, always in black. Sporting harness may be black or brown. Utility harness is for country or trade vehicles."

In regard to horses, Mr. Ryder continues, "For a Park coach or Drag, horses with stylish action are wanted and they are to have good manners, perfectly matched in color

FAMILY CARRIAGE. Typical 4-seater for family use. Note lamp attached to driver's seat. This vehicle might be referred to as a "depot wagon".

EARLY CONVERTIBLE. Appropriately, the folding top was down on this sunny day when this foursome went for a drive.

SUNDAY DRIVE. In 1905 the Lake Shore Drive in Chicago's Lincoln Park was used by promenading pedestrians and horsedrawn carriages.

WITH A FRINGE ON TOP. The canopy top with fringe shades this family driving through the Vermont countryside.

ROOM FOR THE FAMILY. In Studebaker's 1912 catalog a vehicle similar to this one was called a "Canopy Top Suburban Wagon". This carriage seated six adults comfortably.

CLASSY OUTFIT. A man who cared about his horses and was proud of them wanted to have a team well-matched in color and size, such as the team shown in this photo. The carriage's spindle seats also enhance the outfit's appearance.

119

PRIVATE TRANSPORTATION

and stamp. Flashy colors and excessive white markings should be avoided.

"On the Road Coach the horses should be of a similar height and weight. Any color, one gray, or odd color, adds to the sporting look. They should be active, have good conformation. High stepping action is not desired."

This kind of unusual and finicky attention to detail most certainly made for elegant and lovely turnouts.

From the extreme of the fashionable stable with 10 or 15 coach and carriage horses, an array of shiny vehicles and a retinue of coachmen, footmen, grooms, and stable boys, to the man owning one horse and one buggy kept in a small barn behind his house, there were many variations in stable sizes.

Having your own horses did create problems. There were constant chores to do. The animals had to be brushed and curried, fed, and watered and stalls mucked out daily. Should the gentleman who was head of the household not have the time, or desire, or felt it below his dignity to do these things, he had to have hired help. Then there was harness to be cleaned, hay to put in the mow, horses to be shod, vehicles to be washed, axles to be greased and manure to be disposed of.

Whether a man had one horse or a dozen was determined by his financial position and the extent of his interest in horses. If he loved owning horses, all the headaches of maintaining his own stable were worth it. This was also a status symbol in the community.

If these things did not concern him, there was always the livery stable where he could board his horses, or rent a horse and buggy when needed.

When the snow fell, sleighs would be brought out of storage. Bob sleds were the large vehicles and had two sets of runners. The front runners turned under the body,

SIDE SADDLES. Gentlemen rode astride and ladies rode side saddle. Women looked more graceful and feminine on this style of saddle.

STEPPING DOWN. As the lady departs the Cabriolet, she uses the stepping stones that were frequently placed in front of homes. Hitching posts were installed at the curb.

POSTILLION-DRIVEN. Carriages such as this "grand Victoria" became associated with the extremely wealthy. Mrs. August Belmont poses for a photograph in front of her Newport, Rhode Island home "By The Sea". The four attendants wear proper livery. Note the pole on the carriage curves downward, then up. If it were straight it would bang against the right leg of the postillion rider on the near wheel horse.

PRIVATE TRANSPORTATION

making it easier for the team. Cutters were single runner vehicles that usually held two people. There were variations in sleigh designs to fit about anyone's desires.

As a horse trotted over the snowcovered road and the sleigh whispered along behind, it was almost a silent ride. Thus, the sleigh bells were an added accoutrement, not for romantic reasons, but to act as a "horn" to warn pedestrians and other sleighs.

Some bells were fastened to the shafts, or across the hames or back pad. Sometimes bells were mounted on a leather strap that fit around the belly of the horse.

Other equipment used in sleighs were the heavy lap robes, perhaps made from a buffalo hide, foot warmers, hand muffs, and ear muffs. Occasionally, to be dressy, colorful feather plumes were fastened to the dash to add a touch of flash to the outfit.

In this era a man who was looking for a new horse might explain to the horse dealer that he needed a nice quiet animal because on occasion his wife or teenage daughter might drive. The dealer always seemed to have just what was required and assured the buyer the horse was "lady broke". On the other hand, if the customer wanted a team with a little "get-up-and-go" the dealer would trot out a good looking team of blacks and say, "Now, here is a spanking good team—just what you are looking for."

The following story on Milwaukee's Horse and Buggy Days, written by Kirk Bates, appeared in the July 15, 1951, issue of the Milwaukee Journal. Mr. Bates' comments are probably typical of any big city in the country in that period of time.

NEW YORK TRAP. The Elkhart Carriage and Harness Mfg. Co. of Elkhart, Indiana, called this trap stylish and light running in their 1899 catalog. The price was $78.00 C.O.D. or $75.66 if cash was sent with the order. Customers specified whether they wanted the trap equipped with a pole for a team or shafts for a single horse.

Credit: Milwaukee County Historical Society

BRISK PACE. A spanking team moves this brougham at a good trot.

Credit: State Historical Society of Wisconsin

HORSE-DRAWN OUTING. When a group went on a picnic, or flower picking outing, the wagonette provided the necessary seats. In case of rain the side curtains unrolled to protect the riders.

Credit: Author's collection

Milwaukee in the Horse and Buggy Days

Until Electric Streetcars Came in 1890 Transportation Was Slow and Irritating for Most People but the Carriage Owners Made Quite a Showing With Fancy Steeds, Liveried Coachmen

YOU can get an argument from nearly any old-timer on whether Milwaukee, when the city moved under horse power, was a pleasanter place than now. Certainly the town was mellower when people didn't go so fast, so far nor so often.

Milwaukee was a horse town from its founding until shortly before the First World War, for the city was slower than most to take up the automobile.

From 1860 to 1890, long suffering citizens put up with horse cars. These were slow, service was infrequent and transferring from any one of the city's three lines was impossible.

When electric streetcars revolutionized transportation in 1890, many people gave up their driving horses. The aristocrats of Prospect ave. and Grand ave., however, kept stables of the finest steeds and fanciest rigs. Those were the days when Gustave Pabst and L. J. Petit would drive out on the tree lined Whitefish Bay road on a Sunday morning in a fancy trap pulled by a prancing four in hand. They were among the few Milwaukee drivers who affected four in hands, but there were high stepping teams and gaited thoroughbreds aplenty. Some of the fancy carriages were made in Milwaukee by G. W. Ogden & Co., some imported from Europe.

Leading businessmen were driven to their offices in the morning, called for at night by the coachman. Later in the day, madam might summon the carriage for a shopping expedition, and almost always fancy rigs with liveried coachmen and sometimes footmen were lined up at Gimbels and Chapman's.

The larger east side homes kept elaborate carriage houses, with upstairs living quarters for coachmen, footmen and grooms.

Carriages Lined Up at Theater

Perhaps the greatest display of fancy rigs in that day was seen Wednesday nights when the city's German aristocracy attended the Pabst theater. Carriages arrived in a steady stream and when the show was due to be over, the carriages and drivers were lined up for blocks with eagerly pawing, restless horses.

On Sundays the aristocrats frequently took drives. The Whitefish Bay Resort, boasting the world's first Ferris wheel, was a popular spot with the carriage trade summer and winter, for a dinner in the country. Lesser folk could go by boat in the summer—25¢ round trip from the Gimbel dock—or by the "dinky," a narrow gauged steam railroad, at any season.

Livery stables thrived. Boynton's, White's and Hans Berg's were among the best know east side establishments. All had a wide selection of horses, and quite a variety of buggies and carriages. Hacks and open carriages were available for weddings and funerals.

Hacks and Rubberneck Wagons

Transfer of railroad passengers from depot to depot was a big business, so competitive in fact that an ordinance in 1901 fixed fares at $1 a mile for the comparatively short trip. One of the best known hack drivers of the town was Mr. Fog Horn Smith, so known for his tremendous voice. When business was slack he tried to stir up trade by his raucous yelling.

Hack fares were high—roughly $1 a mile for one or two passengers—and not too many could afford to ride in them. Some of the nighttime business of the hack drivers was hauling patrons to and from the River st. red light district. Drunks also were often addicted to hack riding, and many a sport topped off a spree by being driven around the town while he slouched on the cushions with his feet sticking out of the hack window.

Milwaukee's first "rubberneck wagons" were horse drawn tally-hos that met the excursion boats at the Goodrich dock and took passengers for a ride to the interesting points in the "Cream City" (so-named because of the cream colored brick made here). These tally-hos held 26 passengers each. There were six of them, operated by a Chicago company.

On days when the number of passengers on the boats was heavy, or an unusual number wanted to go for the sightseeing trip—price 75¢—some of the ordinary buses of the town would be hastily pressed into service to handle the overflow. Then the six tallyhos and the string of buses would form a lengthy procession through the east side and parts of the city adjacent to downtown.

There were frequent complaints about the spicy lectures the drivers gave in passing the homes of leading citizens.

"And this beautiful brown stone house on the right is the home of Mr. Etc., leading Milwaukee brewer, whose wife, a jealous creature, took a horsewhip to him when she suspected he was having a bit of an affair with a local actress. On the far corner to the left is the home of Mr. Blah, a wealthy meat packer of whom it has been said. etc., etc."

Carriages Decorated With Flowers

A feature of Milwaukee's horse and buggy days at the turn of the century was the flower parades organized by the late Mayor David S. Rose as a part of his "all the time Rosey" program to give the citizenry a continuous circus. The flower parades were huge, elaborate and beautiful.

Owners of fine carriages decorated their rigs and their steeds with a profusion of artificial flowers, achieving unusual and beautiful effects by their color schemes. The end of the parade was brought up by heavy brewery wagons and other drays and trucks similarly decorated.

Mayor Rose, driving his own high stepping team, led the procession between banks of cheering spectators.

Another form of horse drawn amusements popular at the time was fire runs in which fire engines—sometimes fired up and with smoke and sparks pouring from the chimney—were pulled through the streets at breakneck speeds of possibly 12 miles an hour.

The well trained fire horses were kept loose in box stalls with a swinging door which a horse could open by butting with his head. The horses were trained at the sound of the gong to charge from their stalls and take their place before the fire engine.

Specially constructed harnesses suspended above them were quickly dropped on the horses' backs and fastened by two buckles, so that the fire truck was away at a gallop within half a minute after the alarm sounded. To keep the horses in trim, rehearsals were held every afternoon, and fans used to gather in front of the fire stations to see the horses charge out of the stalls when the gong sounded.

The Big Horses of the Breweries

The freight of the town, as well as the people, was hauled by horse power. Drays, trucks and delivery wagons were everywhere, the heavier loads pulled by huge and beautiful draft horses, the finest specimens of which were likely to be owned by the breweries. A fine team of prancing Percherons, handled by a muscular giant in a leather apron, and hauling a wagon piled high with beer barrels, was considered an excellent advertisement for the brewery.

So that was a glimpse of Milwaukee in the good old days of the horse, before automobiles took over. Good old days? Well, maybe, but it is doubtful that even the old-timers with the most nostalgic memories would want them back again.   —KIRK BATES.

KNOWLEDGE IS POWER.

How Owners of Carriages Often Abuse Rather Than Use Them.

In a little booklet, the Muncie Wheel Co., Muncie, Ind., publish a wealth of information to those who buy, sell, or use a buggy, as well as to carriage makers and repairers. Commenting on the way that the owner often uses a buggy, which was in the first instance as near perfection as it could be, they say:

"We are asked frequently, 'How often should one have the tire reset on their wheels?' We would ask such a one, 'How often should you have a horse shod?' You would probably answer at once, 'Whenever the horse needs it.' We give you the same answer with reference to the resetting of your tire. Whenever you take your horse to the blacksmith to have his shoes reset, you do not expect him to do his work for nothing, nor should you expect the carriage manufacturer to reset the tires on your wheels without charge.

"The tires on your wheels when you buy your buggy have, as a rule, been put on in the most careful and mechanical manner. This being the case, the responsibility of the manufacturer of the wheel and the manufacturer of the buggy ceases, so far as keeping the tire tight is concerned. When the buggy is delivered to the purchaser it is then up to the user of it to protect his own property.

"The conditions with which the wheels upon your buggy have to contend while in service make the tires come loose sometimes in thirty days, or they may not come loose in two or three years, but when they do come loose, it is as much the duty of the user and owner of the vehicle to have them reset immediately, at his own cost, as it is to have your horse shod when his shoes are loose.

"The nature of the material of which wheels are constructed (wood) and the nature of the material in the tire on wheels (steel) are directly opposed to each other, and men cannot change them. The nature of the wood is to shrink and grow less in hot, dry weather. The nature of steel is to expand and grow larger in hot, dry weather. This being the case, in a hot, dry time, the tendency of the wood is always to contract and the tendency of the steel is always to expand. The result is loose tires, for it is impossible to keep the tires tight, provided the wheel is given any service during the hot, dry time.

"There is a directly opposite effect in cold, wet, damp weather. The nature of the steel is to contract. In instances of this kind the wheel will take on much more dish than it should have, and if the wheel is permitted to remain with too much dish it very often becomes worthless.

"Now, in both of these instances, the responsibility lies with the user of the vehicle, and he should at once have the tire taken off and reset; if the tire is loose, it should be put back on the wheel tight, and if it is too tight, so that there is too much dish in the wheel, it should be expanded and put back, giving the wheel only the proper dish."

CARRIAGE CARE. Article from the August 1914 issue of *The Blacksmith and Wheelwright* magazine.

GOVERNESS CART. Note the beautifully fashioned wicker basket body, two side seats, upholstered in English corduroy and rear door with a large step under the door. Two sizes were usually available—one for Shetland Ponies, and one for Welsh Ponies. This cart was photographed in 1914 in the town of Fitchburg, Wisconsin.

Credit: Phil Fox Collection

PRESIDENTIAL VISIT. In 1897 President William McKinley (seated in the right rear of carriage) visited one of Troy, New York's, famous manufacturers of shirts and collars and cuffs—Cluett, Coon & Co.

CUSTOM COLOR. This Whitechapel Cart with a tandem hitch had a dark green body with black moldings. The undercarriage could be yellow or vermillion, with black striping as desired

SPIDER PHAETON. These lightweight carriages were ideal for touring behind a spirited team.

City Horses

≫◦ *Boyden Sparks* ◦≪

IN THAT year, 1886, the effect of four years of the National Horse Show had worked a change in the character of the horses that drew New York carriages; it had changed as well the coachmen and grooms. No longer was a mere hired man good enough for such a job.

"You were broke the same as a horse," I was told by John Ryan, who, in 1886, newly landed from Ireland, entered the proud caste of rich men's coachmen by way of the stables of Cornelius Vanderbilt. "Yes, sir, that was my first horse job in this country, with old Cornelius Vanderbilt. The one that's General Vanderbilt now was a boy then, with shoe-button eyes, riding around on a bicycle.

"John Doyle was the coachman. You spell it D-o-y-l-e, not D-i-a-l, as I say it. I started in the stable as a groom, cleaning harness, washing and dusting carriages. They had to be spick-and-span when Doyle came around, or he'd hit you over the head with something. He was a fine fellow, very strict, who wouldn't make free with anybody.

"The stable was in Fifty-eighth Street between Madison and Park, on the south side of the street. We had eight to ten horses in the winter, but in summer at Newport, at The Breakers, we'd have about twenty-five, and when I say 'horse', though I pronounce it with an *a* instead of an *o,* I mean horse. Blooded animals every one, of course.

"I never rode on the box with the Vanderbilts, except at Newport to the races, when we had guests from the city. We all had livery and uniform stable suits. The Vanderbilt livery was maroon with white breeches. The brougham was maroon, too, with red pencil stripes. Boots, of course, were black, with pink tops. There was no cockade on the hat.

"The old Vanderbilt would drive himself mornings in a dog cart, sitting back to back with a second man. I've been that second man, and the instant he stopped, you'd jump down and take

---

LIFE OF A COACHMAN. This story appeared in a 1930 issue of the Saturday Evening Post, recalling an already bygone era. The excerpt is part of an interview with a coachman, which presents a first-hand account of the job.

≫◦◦◦◦◦◯◯◯◦◦◦◦◦≪

---

the horse's head. Starting the instant he had the lines, you'd have to be up on your seat, or he'd go off without you. At the best, he'd be going while you were getting on. 'Twas adventure to be riding in America with a Vanderbilt.

"If you rode as footman in them days, the instant the brougham stopped, you must be down off the box, run around to the rear so as not to frighten the horses. You opened the door and took the carriage robe, being careful to keep your head turned so as not to breathe on the lady. You wouldn't go grab her then, like I've seen chauffeurs do, but you'd put your hand out stiffly and if she wanted assistance she'd lay her hand on your arm.

"In them days, during summer, sometimes at night the stable was cleared out for a dance. Mr. and Mrs. would come down and look things over. Maybe they'd enter the grand march. Plenty of stuff would be sent down from the big house. Nearly every night there was a racket to go to when there was no work to be done.

"You couldn't go to a school a few nights in them days and get a license, as you can now, for an auto. It took years to be a coachman. To be a coachman you had to know things. You weren't really a coachman till you could take a horse's temperature by putting your hand in its mouth, like John Doyle could do."

Losing a Race and Three Hundred Dollars

ONCE a vet was called to see a lame horse when Doyle was away. 'It's his ankle that's twisted,' said the vet. Then Doyle came in. 'The horse is standing on his toe,' he said. 'It's something in his hoof.' He took the shoe off, explored with a knife and found a nail, which he took out with pincers. Then he washed out the hole with spirits of salts. They got a new name for it now, though it's the same stuff when you call it muriatic acid. When we called a vet it was out of friendship. Many a vet, when he was puzzled, would ask a coachman to look at an ailing horse.

"The most of the coachmen were Irish, Scotch-Irish or English, with the Irish in a majority, but all things were done in an English manner. It was John Doyle who got me my place as a coachman with Theodore W. Myers. He was a banker and broker and Comptroller of New York City. I used to ride in them days. Would you think it now that forty years ago I weighed but a

READY FOR THE COLD. Winter equipment for horses included blankets ranging from the elegant ones illustrated, to an old blanket covered over with a piece of canvas to keep out rain and wet snow. The coachmen, in the illustration, are also winter attired with robes and capes.

Credit: Author's Collection

hundred pounds? I rode Tormenter, a Myers horse.

"In one race, neck and neck was a gentleman polo rider I won't name to you. 'Three hundred to you if you lose, Ryan,' he said. I kept riding, and then, through no fault of mine, the horse tripped and the gentleman polo rider came in first. 'How about that three hundred?' I said to him. 'You can go to hell, Ryan,' he said. Afterward, at Staten Island, I brought Tormenter in first, but we'd lost ten pounds of lead out of the saddle, and because we came in light we lost the race.

"They had a gray and a chestnut in the carriage team when I went with them. When that gray knew Mrs. Myers was in the carriage, he would not pull a pound. She wouldn't let you hit him and he knew it. Afterward it was a team of bays that ran away with her. They were green horses, and once the elevated train had frightened them. That was near the stable and she was not inside. One reared and came down on the pole, cracking it. I put a foot on each rump until I could stop them, which was not for several blocks. The next time they ran away she was inside. The horses had grown cold while she was visiting her sister, in Fifth Avenue near Seventy-second Street. When those bays heard the door slam they wanted to get home. They ran from Seventy-second Street to Fifty-ninth. After that I put the reins from plain cheek to lower bar, and I could hold them. While I did it, Mrs. Myers was leaning out of the brougham telling me I'd saved her life."

Beau Brummell on the Box

AFTER eight years with Comptroller Myers I went to Mrs. Margaret Plant, the stepmother of Morton Plant. I was there twenty years, though she was held hard to please and had had five or six coachmen in a short while before I went with her. I think that was when the carriage parade was finest. People used to sit for hours and hours just watching the horses and turnouts going in and out of the park.

"Mrs. Plant was in 586 Fifth Avenue, across the street from Helen Gould. You would go for orders to the mistress every morning at nine, Sundays and all. She'd say: 'I want nothing or, 'Come at three, weather permitting,' or whatever she had on her mind. When she did come out you touched your hat with your finger and there'd be no word from you, but she would examine you from head to foot when she came down the steps. 'Your hat doesn't look so good,' she might say. 'Take it to the cleaners' tomorrow and have it ironed.' Or maybe she'd look twice, and hard the second time, and you'd have to guess what was wrong with you.

"I wore a low-crowned black silk hat summer and winter. Summers my coat was blue broadcloth with silver buttons. The starched white collar lay flat to the coat collar, and there was a white puff tie which had an extra virtue in that it concealed your shirt. Spring and fall the breeches were white buckskin, and cost twenty dollars a pair. The breeches of stockinet for summer cost twelve dollars. Boots, patent leather, tops included, cost twenty-five dollars. The five inch of pink top was detachable and had to be treated with a dressing from England. Fawn gloves for summer were five dollars, and the fur ones for winter cost thirty. The winter overcoat of heavy blue broadcloth with astrakhan collar and military frogs cost one hundred and forty dollars. A coachman's pay was seventy-five dollars a month, and a second man got fifty or sixty.

"It was the custom of most coachmen to part their hair at the back, brushing it forward toward their ears. We did not read newspapers or gossip with one another in front of stores, like the chauffeurs today. We did not smoke or chew or relax at all; and if we were that careful of ourselves and our manners, you can imagine what was done with the horses and carriages.

"If a horse had a white stocking it had to be kept white with bluing and cornmeal. Sometimes it was necessary to keep pants on them to avoid stains. It cost twenty dollars a month to feed each animal the best of oats, timothy hay and bran. Shoeing was another fifteen dollars, including the rubber pads made necessary by the asphalt. Harness cost up to $375 a set. The horses themselves would be fair ones at $1,500 a pair.

"Gibson, the dealer, kept the best sales stable, in Fifty-first Street between Broadway and Eighth Avenue. It's a garage now. When a new pair was needed, Mrs. Plant would sit in her Fifth Avenue window and watch as they were driven past. If they pleased her she'd nod and then take a drive behind them. She'd never see the bill or ask the price. Her secretary would attend to such matters.

"If she went shopping she would hire a vehicle. Her own blue brougham was never taken below Twenty-third Street. The victoria was for fine weather; the brougham for damp. We had a vis-a-vis, which was a four seater with a white canopy over it. Brewster was the best carriage builder, in my judgment, and she would go to his place at Forty-seventh and Fifth Avenue to look them over, but never to a horse yard. You had to bring the horses to her.

"On the box the tops of the hats of the coachman and footman had to be plumb level, and if one was higher than the other, sitting, the smallest had to be built up with a pad. Where will you look today for such care in appointments? Surely not among these things the chauffeurs drive while most of them lounge on their spines and toot their horns without regard for anybody at all. Why, there's few of them would know whether you were going left or right if you signaled with a whip." ☺

COLONEL KANE'S COACH. J. B. Brewster & Co. of New York published this print in 1876, showing one of the coaches they built.

ON AN OUTING. One of many styles of roof seat breaks used for picnics or outings. C-1890.

A FORMAL GATHERING. An example of a park drag.

SLEIGH BELLS RING. On the far left is a body strap that went around the body of the horse. Top left, bells that were fastened to the shafts or pole. At lower left, bells that were attached to the saddle of the harness. Bells were usually brass and plated nickel, gold, or silver, according to the customer's wishes. The better bells were cast brass; less expensive bells were stamped. Bells had more purpose than merely the romantic effect—a horse trotting along a snow-covered street with a sleigh whispering behind made very little noise. The bells signaled oncoming traffic.

GOING FOR A SLEIGH RIDE. With buffalo robe tucked snugly around the young girl, the driver is about ready to depart the porte cochere. One horse is usually enough power for a cutter this size. This gentleman will have an easy time on his trip with a team. C-1900

TYPICAL CUTTER The sleigh bells are fastened to the shafts. Note the heavy lap robe. With a conveyance like this, the lady could go shopping or visiting in 1908.

CLEAR "SLEDDING." If a horse was properly shod, a frozen lake cleared of snow became a fast track. Here Dr. Rod Fox of Madison, Wisconsin, exercises his horse on the ice of Lake Monona around 1910. Frequently trotting races were held on this lake. Note the string of sleigh bells on the horse.

WAITING FOR MR. CASE. The coachman awaits the owner: in this instance, Mr. J. I. Case of Racine, Wisconsin. This bob sleigh is very similar to the drawing from a catalog. The bob runners (2 short sleds) provided a smoother ride over rough and rutted snow-covered roads.

JINGLE BELLS. A single horse could handle a bobsled like this easily, unless the snow was very deep. Note the string of bells on horse.

GROUP TRANSPORTATION. Taking a group to town or church during the winter often meant using a farm bobsled. It wasn't elegant, but it was practical. This photo was taken in Steuben, Wisconsin in 1900.

A BOBSLED FOR A CROWD. A bobsled this size could handle ten to twelve people on each side in seats that ran the length of the body. The rear end was open and equipped with a step for easy access. Runners were five to six feet long to glide the load over the snow. Usually a sleigh of this size had a four, or even six-horse team. It was used as a bus to convey people from the railroad depot to various destinations, or perhaps for a pleasurable sleigh ride.

RUNNING ERRANDS. A bobsled, with a seat for two, had plenty of room for hauling packages and parcels accumulated on a shopping trip to town.

CHAPTER VIII

BUILDING THE VEHICLES

WOODWORKING SHOP. In Milwaukee, the Gridley Dairy Co. manufactured their own milk wagons.

SOME FACTORIES were so enormous they could turn out tens of thousands of vehicles a year. And some only turned out a couple of dozen a year. In between there were scores upon scores of wagon works of various capacities which specialized in certain kinds of horse-drawn equipment.

There were companies that, because of distinctive technical requirements, produced only certain kinds of rolling stock, such as fire engines or massively built circus wagons. Others featured wagons used for freighting and dray work. Here again were specialists who, perhaps, had the best line of roll wagons for the brewers or produced earth- and sand-hauling wagons for construction work. Some factories featured delivery wagons and many only produced pleasure vehicles. There were specialists who catered to those who wanted elegantly appointed carriages.

Just about any town or city had one or more factories producing the run-of-the-mill requirements. This was possible because there were so many factories around the country that produced various parts. Some offered bent felloes, carriage seats, wheels, lamps, axles or hardware. In other words, the small operators did not necessarily build a buggy, or dray wagon, from "scratch".

Then there were the GMs, Fords and Chryslers of the era. Studebaker Bros. of South Bend, Indiana, produced 75,000 horse-drawn ve-

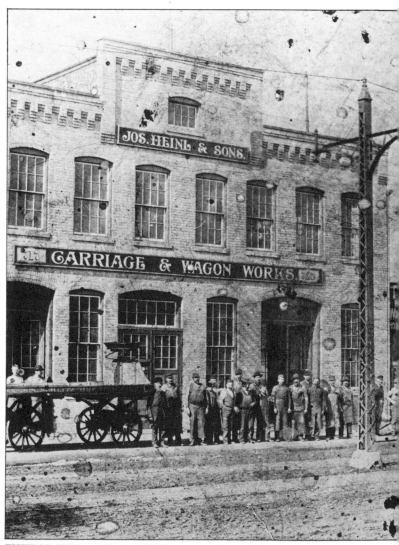

THEY'LL KEEP YOU ROLLING. Jos. Heinl & Sons operated their wagon works at the corner of Ogden Ave. and North Water St. in Milwaukee. According to the sign they offered to set loose tires while a man waited. This was an important service, as if a tire came off, a wheel would be ruined quickly if the driver kept going.

DIVERSIFIED BUSINESS. Cheney & Ranken Carriage, Sleigh and Harness Emporium in Troy, New York, sold robes, lap blankets, horse clothing, buggies, surries and sleighs of all descriptions, single and double harness, and ladies' and gents' riding bridles and saddles. As a sideline, one of the partners was a "Land Plaster."

90,000 WAGONS A YEAR. Big plants such as the Kentucky Wagon Manufacturing Co. in Louisville produced 90,000 wagons a year.

QUALITY PRODUCTS. The Beggs Company specialized in circus and wild west show wagons, requiring extra strength built into the vehicles.

BUILDING THE VEHICLES

hicles in 1895. (By 1919 Studebaker went out of this business and produced only automobiles.) Elkhart Carriage and Harness Co. of Elkhart, Indiana, manufactured 10,000 vehicles a year in 1896, and some of their buggies were priced as low as $49.50. Luth Carriage Co. of Cincinnati turned out 175 vehicles a day in 1913. And there were others that produced horse-drawn equipment on a grand volume.

Mass production brought prices down. In 1890 Murray Mfg. Co. had a carriage priced at $59.50. Eureka Carriage Co. offered a buggy at $37.50. About 10 years later the Barnett Carriage Co. had a buggy priced at $29.90, and a surrey at $59.90.

Then, of course, one could purchase vehicles from the elaborate catalogs of Montgomery Ward and Sears, Roebuck & Co. These mail order companies supplied the farm trade, in particular, and offered all kinds of wagons and carriages, farm implements, harness and various parts and pieces.

It was big business. The Carriage Association estimated that in 1900 907,483 pleasure vehicles and 118,221 sleighs were built.

The wagon and carriage manufacturers employed a large array of talented craftsmen. Blacksmiths were needed for the iron work. There were woodworkers, assemblers, painters, polishers, and those who rubbed the vehicle to a mirror-like finish. Then there were the stripers and the trimmers who handled the upholstering chores.

The catalogs published by the larger companies illustrated their wares in magnificent steel engravings. The quantity of variations, styles and models offered was astounding. Henry Hooker & Co. of New Haven, Connecticut, for instance, brought out 175 different models of Phaetons.

The builders seemed to cater to every

OLD RELIABLE. The Springfield Wagon Co of Springfield, Missouri, used this dramatic drawing in their advertising in 1910.

ANOTHER MANUFACTURER. This drawing appeared in Parlin & Arendorff Co.'s 1915 catalog.

WAGON WORKS. Haekler & Habhegger Carriage & Wagon Manufactory was another Milwaukee producer of vehicles. At the turn of the century they turned out a fine line of industrial, commercial and private vehicles, specializing in various kinds of beer wagons for the city's breweries.

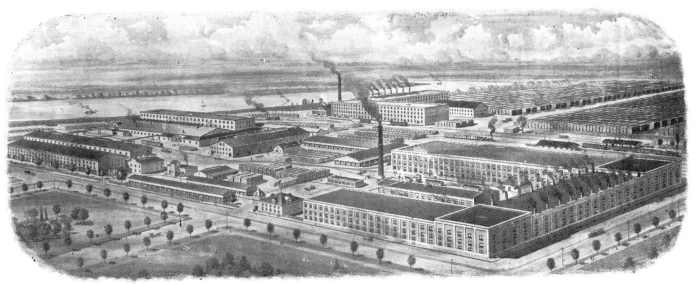

IN DEMAND. Owensboro was a big manufacturer and produced quality wagons. They had an enormous plant - C-1915.

Office employees each day will fill lamps, clean chimneys and trim wicks. Wash windows once a week.

• Each clerk will bring in a bucket of water and a scuttle of coal for the day's business.

• Make your pens carefully. You may whittle nibs to your individual taste.

• Men employees will be given an evening off each week for courting purposes, or two evenings a week if they go regularly to church.

• After 13 hours of labor in the office, the employee should spend the remaining time reading good books.

• Every employee should lay aside from each pay day a goodly sum of his earnings for his benefit during his declining years so that he will not become a burden on society.

• Any employee who smokes Spanish cigars, uses liquor in any form, or frequents pool and public halls or gets shaved in a barber shop, will give good reason to suspect his worth, intentions, integrity and honesty.

RULES AND REGULATIONS. Henry W. Meyer, in his fascinating book, "Memories of the Buggy Days", lists a set of regulations the owner of a Cincinnati carriage works posted for his office help in 1872

The U.S. Census Bureau in 1890 listed 80 carriage and vehicle accessory manufacturers in Cincinnati employing 9,000 workers and producing 130,000 vehicles. By 1911 the number of manufacturers was down to 32.

BUY YOUR CARRIAGE HERE. Davis & Vannier in Troy, New York, were sales agents for pleasure and business vehicles. Most towns of any size had one or more companies that made the usual vehicles, and nearly every town had a dealer. But if not, there was always a Montgomery Ward, or Sears, Roebuck.

135

BUILDING THE VEHICLES

whim, with an enormous number of vehicle types and many style variations. Here are a few:

Gig	Caleche
Skeleton Gig	Omnibus
Very Spicey Gig	Sulky
Paris Lady's Chaise	Roadster
Village Cart	Roadwagon
Mail Phaeton	Surrey
Stanhope Phaeton	Rockaway
Spider Phaeton	Wagonette
Dog Cart	Depot Wagon
Cabriolet	Runabout
Hansom Cab	Buckboard
Breaking Cart	Errand Wagon
Jaunting Cart	Vis-a-Vis
Lady's Brougham	Lady's Country Trap
Bachelor's Brougham	Spring Buggy
Landau	Panel Seat Democrat

GABRIEL STREICH, OSHKOSH, WIS.

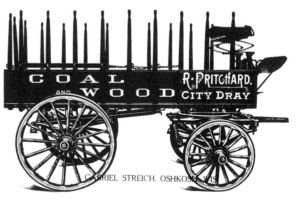

HEAVY-DUTY. A. Streich & Bro. Co. of Oshkosh, Wisconsin, manufactured a line of heavy-duty wagons, including ice delivery vehicles.

WAGONS FOR HEAVY HAULING. These four illustrations are from the catalog (C-1910) of Gabriel Streich Co., Oshkosh, Wisconsin. They manufactured an extra durable and well-built line of dray wagons.

CHOOSE FROM 256 STYLES! Dog carts, goat carts, cute pony phaetons, governess carts, runabouts, and a classy buggy finished in natural wood were offered by the Michigan Buggy Co. of Kalamazoo, Michigan, in this ad in the June 22, 1910, issue of Breeder's Gazette.

No. 149—New Autostyle Buggy.
Twin Auto-Belgian type seat with large, round corners and beautiful curves. Handsome, stylish, comfortable. All wrought gear and best second-growth hickory gear woods. Axles arch, high arch or very high arch as desired. Soft, easy-riding springs, oil tempered. 2,500-mile, long-distance, dust-proof axles. Our A-grade select hickory wheels and high-bend select white hickory shafts with 36-inch leather tips. Latest auto design top with "automobile" leather. Painting and trimming optional.

No. 60—Two-in-One Auto-Seat Buggy.
The quick-shift top can be detached in 30 seconds, leaving handsome and stylish runabout. Great buggy for liveries and for use where it is impossible to keep both a top and an open buggy. Best second-growth frame and hardwood bottom boards will give unlimited service.

No. 329—The New Yorker.
Entire job is constructed of best and most expensive material obtainable. High, square arch axles; special hand-forged wrought steel braces; special stop circle; oil-tempered open-head springs, rubber, with brass bearing; 1⅝-inch Kelly Springfield tires; hand-sewed, hair-stuffed cushions; thick, velvet padded carpet; grain leather dash.

No. 40—Pony Runabout.
Built to fit ponies of any size. Full wrought gear and best hardwood frame construction. Banded wood hub wheels. Heavy broadcloth, whipcord or leather trimming. Painting optional.

No. 277½—Pony Phaeton.
Body constructed of finest imported German reed woven on hand-forged wrought iron frame. Stylish, handsome, durable, and like the rest of our pony work, a job in which grown-ups as well as children can ride with comfort.

Kalamazoo Holdfast Stable and Storm Blankets are made by us complete in our own mills. We make our own kersey, kersey lining, etc. Dependable blankets, the very best that skill, money and knowledge of the business can produce.

No. 363—Finished in Natural Wood
both inside and out. New style seat; double rails framed with mortised spindles. Forged steel mountings. Best easy-riding springs.

Buggy Buyers, Don't Be Deceived!

When a price is quoted you on a buggy from the factory direct, remember—**the freight is seldom included.**

Often the freight alone makes the difference between a low price and a high price.

When you buy a Reliable Michigan Buggy of your local vehicle merchant, **you have no freight to pay.**

Quality for quality, the Reliable Michigan dealer in your home town can quote you just as low prices as anybody can by mail.

And your local merchant—a man whom you know—a business man of standing in your own community—is **always there to make good any statement** he makes concerning our Reliable Michigan vehicle—or that you read in our advertisements.

You have a claim on us not only for 30 days or a year, but **at any time** after you buy the vehicle, if it develops a defect in workmanship or material.

You do not buy a vehicle "sight unseen" when you buy of your local Reliable Michigan merchant. Go to his salesroom and examine on his floor the splendid types of

The Reliable Michigan Line of Buggies and Pleasure Vehicles

You can see for yourself the fashionable styles—right down to the minute—the perfect finish that distinguishes the Reliable Michigan vehicle. You can see the exclusive points of construction that make the Reliable Michigan the most durable in the world by the actual test of time.

256 Styles to Choose From

We make 256 styles of buggies and pleasure vehicles—every approved type, and scores of our own exclusive design. So well equipped is our big factory to turn out special patterns that we can easily furnish you, through your dealer, a vehicle built to your own order.

Every Reliable Michigan dealer has our big new catalog, filled from cover to cover with illustrations from photographs, with descriptions of our entire line—256 models to choose from.

Your dealer can obtain any one of these vehicles in the quickest possible time. The Michigan Buggy Company has established a record for quick shipments.

Leaders for 26 years

The Reliable Michigan vehicles have been leaders for 26 years.

We make 35,000 vehicles a year—**more exclusively pleasure vehicles than any other firm in the world.** Why is this? It is because Reliable Michigan vehicles have stood the test for more than a quarter of a century. There are Reliable Michigan Buggies of the first year's output in actual use today—some in daily use **in liveries** that were sold over 15 years ago.

We Mill Our Own Lumber

We buy our lumber as it stands in the trees and mill it in the forest—hickory—poplar—maple—ash—every stick of timber we use. The selection is made by a master lumberman, who accepts only the high-grade goods used in the Michigan line.

Our lumber is air seasoned, under cover, protected from the weather, from 12 to 20 months

before using. After it is air-dried we always kiln dry it to avoid any possible chance of shrinkage.

Although vehicles are finished in our big factory on an average of one every five minutes, it is three years from tree to finished vehicle—so thorough are our methods.

We build every vehicle, from the ground up, **in our own shops.** Many of our best workmen have been with us more than 25 years.

We Set Our Tires "The Good Old-Fashioned Way"

Every tire is fitted to its own wheel, welded by a ponderous machine, built especially for the purpose, heated by a specially constructed oven, whereby the tire is made to revolve, insuring even heating and even contraction. By this method, and no other, can good results be obtained—no "Cold Tire Setting." We turn out over 100,000 wheels every year—yet we received not one tire complaint last year. In Reliable Michigan vehicles so closely are iron and wood glued and screwed that it is impossible for water to get between, no matter how much washing is done.

Painting Takes 100 Days

Every vehicle receives at least four coats of body filler and a rub-down with pumice stone, three coats of color and a rub-down with pumice stone, a coat of varnish and a rub-down with pumice stone. The process takes from 75 to 100 days, according to the vehicle.

If you do not find Michigan vehicles in your town, send us your buggy dealer's name, and we will send you our handsomely illustrated Catalogue "A" showing our line.

We also make the famous Tony Pony line of vehicles for children. We send pony, harness and cart complete. Our ponies are all thoroughbred Shetlands, gentle and city-broken. The carts can't tip over.

In writing be sure to state whether you want Catalogue "A" on buggies and pleasure vehicles, or Booklet "B" on the **Tony Pony** line.

MICHIGAN BUGGY CO.

179 Office Building   Kalamazoo, Mich.

The Holdfast Storm and Stable Blankets are manufactured by us. The same assurance of quality goes with these blankets as with the Reliable Michigan Vehicles.

No. 367—Belgian Auto Seat Surrey.
Auto top with brass nuts, knobs and front moulding. Best heavy broadcloth or M. B. leather trimming. Select hickory gear, wheels and shafts. Painting as desired.

No. 151—Special Belgian Auto Seat Buggy.
This is just like the No. 149 New Autostyle buggy shown above except that it has straight instead of divided back. "Auto" top, full wrought gear. Finish—the best.

No. 362—Twin Auto Seat Buggy.
Solid bent panel seat, extra large and roomy. Best hickory gear woods, and all forgings best Norway iron. New arch axles. Soft, easy-riding springs, **oil tempered.** Painting and trimming optional.

No. 269—Pony Trap.
The handsomest pony trap ever built. Hardwood body, with panels built of finest German imported reed. Can be used either all facing front or dos-a-dos. We built this trap to fit ponies of all sizes from 43 to 55 inches.

Who Wants Bonnie Boy?
This shows one of the ponies from our herd of over 200 of the finest Shetlands in the country. The cart is the best style built for children's use —the Governess Cart. It is stylish and easy riding. Almost any number of children can pile into one of them and ride with comfort. You can't tip one over. Painting and trimming optional.

No. 20—Speed Sleigh.
A sample of our line of 40 business and pleasure sleighs. We use in their manufacture strictly air-seasoned and bone-dry lumber, single-ply panels and dashes. Our gears are heavily ironed and braced and nothing but Norway iron bolts and rivets are used. Our cushions and backs are padded with curled hair or cotton, own excelsior, and are trimmed in velour, plush or broadcloth. Each cutter is subjected to a most careful inspection at all stages of its construction.

No. 252—Dog or Goat Cart.
28-inch wheels, half oval tires; shaved spokes. Bent white ash shafts and bars bolted and riveted together with wrought iron braces. Single plate special steel spring. All iron work XC plated. Nice upholstered panel seat. Finished in natural wood, or vermilion.

NOTE EVERY POINT OF DIFFERENCE IN THESE TWO VEHICLES AND THEN ANSWER THIS QUESTION

"Which one of these two vehicles, all other things being equal, will give the most Convenience, Comfort and Health?"

If you will do this, carefully considering the points of difference we have brought out in the picture, we have no doubt of the result. You will say:

The Cozy Cab

It is really much more handsome than the old style buggy

We challenge successful contradiction of our statement that no vehicle having storm protection features made in America, approaches the Cozy Cab in soundness of selected materials and thoroughness in putting these together. It is built for hardest service on roughest roads and will outlast a number of factory-made vehicles. And yet it is easy on the horse—lightest draft vehicle in its class. Those who use it most like it best.

But the Cozy Cab is not only a handsome, durable, light-draft, fair weather vehicle. It provides ever-ready, instantly-adjusted protection from rain, snow, sleet, wind, dust or mud. Your protection is not an attachment, but is a built-in part of the vehicle itself. Three one-hand movements open or close it, and your storm protection is never missing when needed, as storm apron and side curtains so often are. Ventilation is perfect.

Send for Catalog—Study it—then Try the Cozy Cab 30 Days before Paying for it

Our Try-Before-You-Buy Plan makes it unnecessary for you to risk a cent on the Cozy Cab. Send for catalog and then test the vehicle by using it before you pay for it. If you are not satisfied with it you need not pay.

FOUTS & HUNTER CO.
No. 77   South Third Street, Terre Haute, Ind.

FANDANGO 143

THE COZY CAB BUGGY was "cool on a hot day, warm on a cold day, and stormproof on a wet day", claimed the manufacturer in this 1909 advertisement.

138

"OwensborO" Platform Spring Dray

"CAPABLE OF THE HARDEST SERVICE." Different models were available to haul loads of 1,800 lbs. to 6,500 lbs.

Studebaker Delivery Wagons

Three Spring Cut-under Panel Side Camel Back Delivery Wagon

"ONE OF THE MOST ATTRACTIVE IN OUR LINE." This 1912 Studebaker delivery wagon was used by laundries, department stores and grocers.

"OwensborO" Platform Spring Dray

"HEAVILY IRONED THROUGHOUT." Owensboro Wagon Co. of Owensboro, Kentucky, had a good line of heavy commercial and industrial freight wagons in various capacities and at various prices.

POPULAR WAGON. Streich's dump wagon was very widely used. The automatic dump bottom was a big feature.

SURREY, BUGGY OR WAGON. Three popular vehicles offered by Parlin & Orendorff Co. of Canton, Illinois in their 1915 catalog.

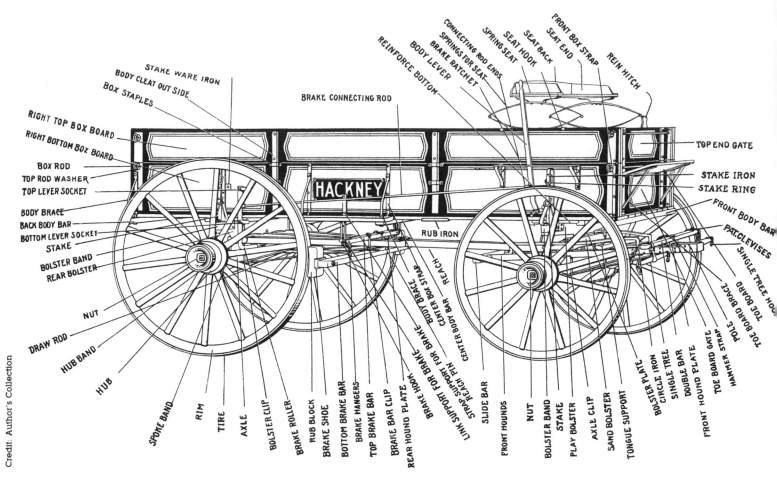

ANATOMY OF A WAGON. This diagram points out the various parts of a wagon, to make ordering replacement parts easier.

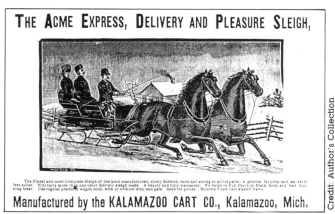

"A BEAUTY." In 1889 the Farm Implement News carried this ad for a line of sleighs with black bodies and red running gear.

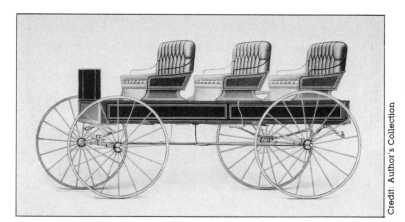

VERSATILE WAGON. The seats were removable in this Studebaker wagon so it could be used for hauling. As extras, the company offered canopy, extension, or stage tops. C-1912.

We Want to Hand You Our Big Book for 1909

SPLIT HICKORY

And Pay the Postage All Free

"I save you $40 on this Carriage"
Split Hickory Light Carriage—Popular for Family Use.
See Page 111 in My Free Catalog.

"I save you $25 on this Auto-Seat Runabout."
Split Hickory Special Auto-Seat Runabout.
See Page 34 in My Free Catalog.

"I save you $30 on this Auto-Seat Buggy."
Split Hickory Square-Deal Auto-Seat Buggy.
See Page 58 in My Free Catalog.

"I save you $20 to $30 on this Cpen Wagon."
Split Hickory Half-Platform Open Wagon.
See Page 115 in My Free Catalog.

Don't Miss It

Don't wait another day or week to write me for this book. You know about our plan, direct to you at lowest prices, and about Split Hickory quality. But you haven't seen our Big, New Style Book for 1909 yet, and I want every possible buggy buyer to be sure to see it before buying any kind of a vehicle or high-grade harness. Just write a line to me today.

Get My Prices and Big Book of 1909 On

Split Hickory Vehicles

Save $26.50 and Up and Take 30 Days' Free Road Test

We make to order 125 styles of Split Hickory Vehicles, including all styles of Top Buggies, Automobile Seat, Two-In-One Buggies, Handsome Runabouts with Fancy Seats, Regular Seats and Automobile Seats, Phaetons, Carriages, Surreys, Spring Wagons, Harness. The Split Hickory Buggy shown at the right here made to your order for $26.50 less than your home dealer's price for anything like the value.

2 Years' Guarantee

NOTE—Celebrated, Sheldon, Genuine, French Point Automobile Springs used on all Split Hickory Vehicles, making them positively the easiest riding buggies on the market.

You'll see all these styles illustrated and described in the best, biggest and most beautiful buggy book ever published. It is our this year's style book. Describes our complete line—the greatest buggy values ever offered. Send for it today. We pay the postage.

Send For Big Free Book Today

"I save you $40 on this Auto-Seat Surrey."
Split Hickory Auto-Seat Surrey.
See Page 98 in My Free Catalog.

"I save you $30 to $35 on this Auto-Seat Buggy."
Split Hickory Special Auto-Seat Buggy.
See Page 79 in My Free Catalog.

"I Save You $30 to $40 on this Delivery Wagon."
Split Hickory Cut-Under Delivery Wagon.
See Page 119 in My Free Catalog.

"I save you $26.50 and up on this Top Buggy."
Split Hickory Special Top Buggy—Over 47,000 in Use.
See Page 77 in My Free Catalog.

H. C. Phelps, Pres., Ohio Carriage Mfg. Co., Station 7, Columbus, Ohio

DON'T MISS IT!! Advertisement in a 1909 Breeder's Gazette magazine. The ad states that the company manufactures 125 styles of vehicles.

Credit: Author's Collection

OPEN SLEIGHS. Cutters offered by Sears, Roebuck & Co. in 1900 could be purchased from $9.50 to $18.50 each.

Credit: Author's Collection

WHEELS OF EVERY DESCRIPTION. Certain companies specialized in the manufacture of wheels. These ads show how they advertised their products.

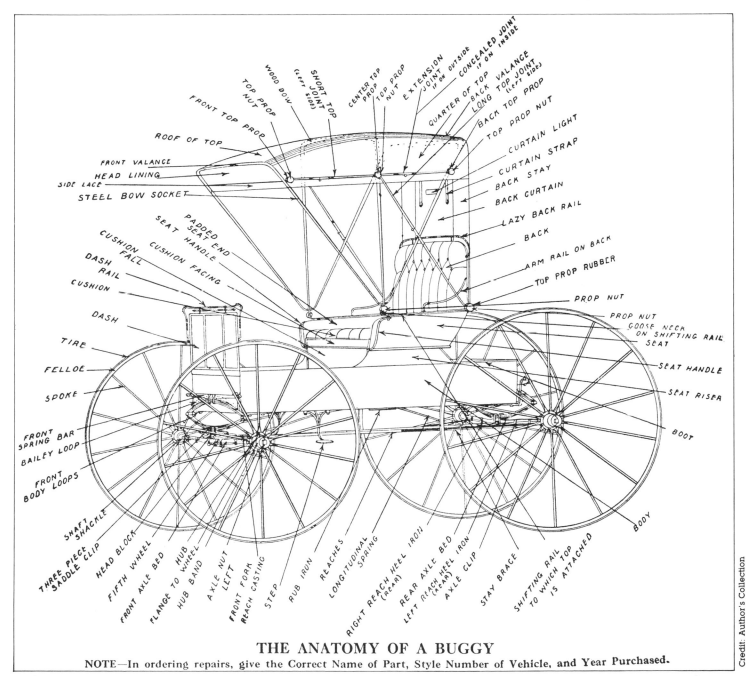

THE ANATOMY OF A BUGGY

NOTE—In ordering repairs, give the Correct Name of Part, Style Number of Vehicle, and Year Purchased.

EVERY PART LABELED. This diagram in the 1915 catalog of the Parlin & Orendorff Co. of Canton, Illinois, made ordering parts easier.

WHILE THEY LAST

The High Class Undertakers' and Livery Mens' Equipment Shown Below, from Actual Photographs Only Recently Taken, Are Offered at the Net Prices Quoted F. O. B. Chicago, SUBJECT TO PRIOR SALE

All the Jobs Shown Have Been Used In the Best Class of Work In Chicago and Represent Extraordinary Values at the Low Prices Offered. We Advise Prompt Action if Interested, as "First Come, First Served. Reservations Ordered by Wire Must Be Followed Immediately With Part Cash Deposit to Secure Job, Deposit Will Be Returned Promptly if Job Desired Has Been Previously Sold.

CLEARANCE SALE. It is very probable that the Funeral Director, Chas. P. Bradley of Chicago put together this flyer to sell all of his horse-drawn vehicles because he was changing to motorized equipment.

No. 87—Stoltze Brougham. Full Size, Rubber Tired, Leather Lined to ceiling. Newly Painted. Upholstery like new. This job is in best of condition, practically as good as new. Worth many times the price.
PRICE $250.00

No. 82 Combination Funeral Coach and Child's Hearse. Top cut shows job as regulation funeral coach; bottom cut as Combination Child's Hearse, taking 4 foot 6 inch Casket. Change easily made in less than five minutes time. Job is Rubber Tired, Leather Lined, with Ribbon Satin Top and in fine condition throughout. A handy, useful job for any undertaker.
PRICE $175.00

No. 73—Cunningham Brougham. Full Size, Rubber Tired, Leather Lined, Newly Painted. Upholstery like new. Has Satin Top. A dandy job worth many times the price.
PRICE $200.00

No. 91—Stoltze Black Casket Wagon. Rubber Tired. Newly Painted, and in as good condition as new. Pole only.
PRICE $250.00

No. 85—Gray Casket Wagon made by Michigan Coach & Hearse Co. Rubber Tired, Newly Painted, good condition. Has both pole and shafts.
PRICE $225.00

No. 78 Cunningham Black Hearse. Rubber Tired. Newly Painted. Plating like new. Has Patent Spring Axles and is very easy running. A stylish job in grand condition and a big bargain.
PRICE $500.00

No. 84—U. S. Gray Hearse. Rubber Tired, Newly Painted and in real good condition. This is a fine job and well worth the money photograph does not do it justice.
PRICE $500.00

No. 16—Cunningham Black Hearse. Rubber Tired, Newly Painted, very light running. Cut does not do it justice. A real bargain at the price offered.
PRICE $350.00

No. 89—Leather Head Landau. Rubber Tired. Leather Lined. Newly Painted. One of the best closed funeral or open pleasure carriages ever put on the market.
PRICE $75.00

No. 92—U. S. Black Casket Wagon. Rubber Tired, Newly Painted, and in best of condition. Has full glass sides with black cloth rays inside, which do not show in photograph on account of reflection. Rays are fitted on frame, which is easily removed and job then used as flower wagon. Flower display rack also goes with job; Wagon has regulation Hearse table bottom, with rollers, pins, etc. A real good job.
PRICE $300.00

No. 80—Stoltze Brougham. Full Size, Leather Lined to Ceiling, Newly Painted. Upholstering Like New. This job is practically as good as new and cannot be duplicated for many times the price.
PRICE $250.00

No. 59 Stoltze Brougham. Full Size. Rubber Tired. Cloth Lined. Newly Painted. A good looking job in good condition.
PRICE $125.00

No. 52 Stoltze Five Glass Landau. Rubber Tired, Cloth Lined, in excellent condition.
PRICE $60.00

No. 68 Seiver & Erdman Brougham. Rubber Tired, Leather Lined to Ceiling, Newly Painted, Full Size.
PRICE $125.00

No. 70—Five Glass Landau. Rubber Tired. Cloth Lined. Newly Painted. Full Size. Very Light Running and ready to go to work.
PRICE $45.00

No. 69—Cunningham Brougham. Full Size, Rubber Tired, Leather Lined, Light Cloth Top, Newly Painted. Good Condition.
PRICE $125.

No. 50 Cunningham Five Glass Landau. Rubber Tired, Cloth Lined good serviceable condition.
PRICE $50.00

CHAS. P. BRADLEY, Manager

JAMES BRADLEY & SONS, FUNERAL DIRECTORS

1820 West Harrison Street

BELL PHONE SEELE 463   CHICAGO, ILL.

Credit: Author's Collection

144

THE CROW BAR

IRONED WAGON WOODSTOCK

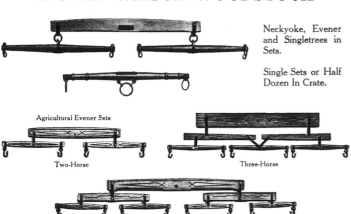

Neckyoke, Evener and Singletrees in Sets.

Single Sets or Half Dozen In Crate.

Agricultural Evener Sets

Two-Horse

Three-Horse

Four-Horse

We Manufacture a full line of Wagon Hardware of which the following is a part

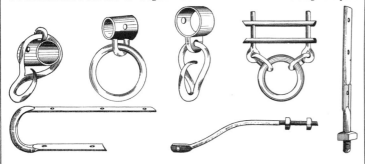

Minneapolis IRON STORE Company
Minneapolis

Credit: Author's Collection

HARDWARE. The Minneapolis Iron Store Co. sold wagon builders and dealers a full line of components.

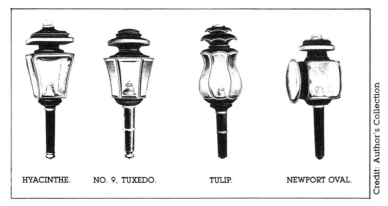

HYACINTHE. NO. 9, TUXEDO. TULIP. NEWPORT OVAL.

Credit: Author's Collection

CARRIAGE LAMPS. From the 1892 catalog of C. Cowles & Co., New Haven, Connecticut.

WAGON IRONS. The Eberhard Manufacturing Co. of Cleveland furnished drop forged and malleable iron castings used in wagons and carriages. This ad, that appeared in the "Blacksmith & Wheelwright" in January 1911 illustrates some of the iron pieces. These and others this company furnished carried very distinct descriptive names, such as "fifth wheels", "doubletree clips", "platform bolster plate", "neck yoke ferrule", "whiffletree cockeye", "whip socket", "sleigh iron", "spring shackels", "felloe plate", "hound clip tie", "pole pin holder", "state pocket", "toe rail", "lazyback iron", "pole crab", "wear iron", and "front perch" iron.

The Whip Industry.

There are fifty-seven establishments in the United States manufacturing whips and parts of whips of rattan, rawhide, reed, whalebone, wood and iron. The fifteen hundred and forty-six wage-earners in this industry are paid annually $703,527, turning out a product valued at $3,948,643 from raw material valued at $1,594,743. In addition to the value of the whip product of the country produced by manufacturers of whips solely, a number of establishments engaged primarily in the manufacture of other commodities reports whips to the value of $114,306, making the total annual value of this product for the United States more than $4,000,000. In spite of the great number of automobiles in use, the whip industry, according to the last census, in five years' time increased the value of its products more than 25 per cent., although the actual number of individual establishments decreased during the same period. Slightly more than half of all the whip manufacturers of the United States are located in Massachusetts, where there are employed 65 per cent. of all the wage-earners engaged in the industry, their product representing more than 70 per cent. of the entire valuation of the whips manufactured in this country.

Credit: Author's Collection

THE STATE OF THE WHIP INDUSTRY. Whips were a small but important facet of the horse industry. This article appeared in the August 1914 issue of the "Blacksmith & Wheelwright" magazine.

Credit: Author's Collection

No. 244. BODY AND SEATS

Price, no seats ....$5.75 and up

Price with seats.......... 7.75

STICK SEAT BODY
$5.90 and up

PIANO BODY

With seat.................. $3.85 and up
Without seat.............. 2.95 "

AUTO SEAT

Untrimmed..$2.65 and up
Trimmed...... 7.48 "

ROUND CORNER SEAT

Untrimmed..$2.90 and up
Trimmed...... 6.05 "

PANEL SEAT

Untrimmed..$0.75 and up
Trimmed...... 3.85 "

BUGGY GEARS

Each.................................$5.90
With Wheels and Shafts...$14.40 and up

Our Profit Sharing Plan
Special Inducements

On orders of $25.00 or over we allow a 3% cash discount and ½ freight charges if you live in Me., N. H., Vt., Mass., R. I., Ct., N. Y., N. J., Del., Va., W. Va., Md., Pa., O., Ind., Ky., Tenn., Ills. or Mich.

All other States ¼ freight charges and 3% cash discount.

On orders of $40.00 or over, we allow a 4% cash discount and full freight charges if you live in Me., N. H., Vt., Mass., R. I., Ct., N. Y., N. J., Del., Va., W. Va., Md., Pa., O., Ind., Ky., Tenn., Ills. or Mich.

All other States ¼ freight charges and 4% cash discount.

On orders of $75.00 or over, we allow a 5% cash discount and full freight charges if you live in Me., N. H., Vt., Mass., R. I., Ct., N. Y., N. J., Del., Va., W. Va., Md., Pa., Ohio, Ind., Ky., Tenn., Ills. or Mich.

All other States ¼ freight charges and 5% cash discount.

TOP SETS

Seat ....$0.95 and up
Cushion................. 1.50 "
Back...................... 2.75 "
Top....................... 5.75 "
$10.95

AUTO SEATS—Top Seats

Seat ....$2.65 and up
Cushion and back... 4.55 "
Top .... 7.75 "
$14.95

POLES

In the white.......................$2.75
Painted 3.60

DOUBLE TREES WITH SINGLE TREES

Buggy and Surrey size.....................Each 75c

BENT REACHES

Assorted sizes .... 6 for 60c

STRAIGHT REACHES

Assorted sizes........... ....10 for 60c

POLES

Buggy or Surrey size ..... Per bundle, $6.95

BUGGY AND WAGON AXLES

50 cents and up
We carry all sizes.

AXLE BEDS

Assorted sizes—drop.............14 for 60c
Assorted sizes—arch.................8 for 60c

SHAFTS   Per Bdl.

Buggy size.................. ....XX, $5.95
Surrey size..............................XX, 6.75

SPRING BARS

Assorted sizes ....14 for 60c

BAILEY BODY LOOPS

30 cents and up

RIMS

We carry all sizes.

Plain Rims........$1.15 and up
Beveled Rims.... 1.40 "
Bored and
Rounded Rims 1.60 "
5 pieces bored and rounded rims, assorted sizes, for 60c

BUGGY AND SURREY SINGLE TREES

Assorted ....6 for 60c

SHAFTS

Buggy size, white...$1.80
" painted... 2.55

LIGHT WHEELS

With Steel Tire on
$5.90 and up
With Rubber Tire on
$12.55 and up
We make all sizes.

BODY LOOPS

2 Sets (8 pieces) for 65c

BUGGY SPOKES

Per 100
1 1/16 in., Sar. Pat........$2.95
1 in., Sar. Pat............. 1.90
1 1/16 in., C. B................. 2.25
1 7/16 in., Sar. Pat........ 3.50
15/16 in., Warner Pat. ... 1.25

Ford Tops............$19.95
Ford Roofs... 7.95

Elliptic Springs
75c and up

DASHES

20 in. drill $0.20
18 in. leather......... 0.45
22 in. leather......... 1.18
25 in. leather......... 0.95

WOOD HUB SPOKES

Per set
2 in......................$2.00
2¼ in...................... 2.30
2¾ in...................... 1.80
2½ in...................... 2.75

Send for our large CATALOGUE and learn how we do it.

Adjust your wants to the sizes and styles we furnish at the above prices and save money.

B. WHEEL, TOP & HDWE. CO.

1104 Sycamore Street   CINCINNATI, OHIO

HEAD BLOCKS

Assorted sizes,
8 for 60c

WHEEL HEAD RIVETS

Assorted sizes
12 lbs. for...............65c

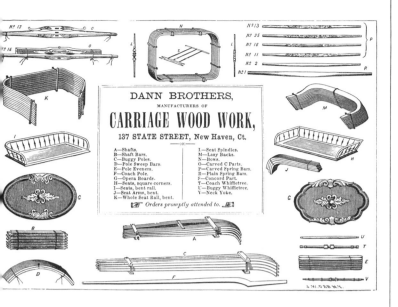

DANN BROTHERS,
MANUFACTURERS OF
CARRIAGE WOOD WORK,
137 STATE STREET, New Haven, Ct.

A—Shafts.
B—Shaft Bars.
C—Buggy Poles.
D—Pole Sweep Bars.
E—Pole Eveners.
F—Coach Pole.
G—Opera Boards.
H—Seats, square corners.
I—Seats, bent rail.
J—Seat Arms, bent.
K—Whole Seat Rail, bent.
L—Seat Spindles.
M—Lazy Backs.
N—Bows.
O—Carved C Parts.
P—Carved Spring Bars.
R—Plain Spring Bars.
S—Concord Part.
T—Coach Whiffletree.
U—Buggy Whiffletree.
V—Neck Yoke.

☞ Orders promptly attended to.

SPARE PARTS. Just about any blacksmith shop, wagon and carriage repair shop or dealer, could purchase parts for vehicle repair. These ads cover springs, axles, wood items, seats, tops, spokes, and dashboards.

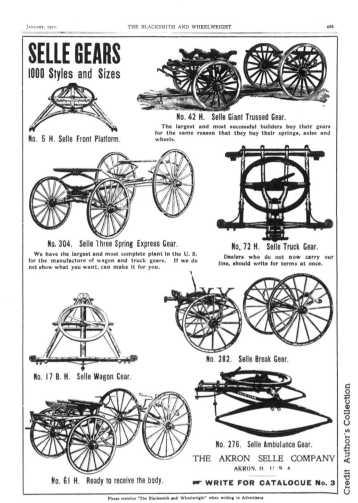

SELLE GEARS
1000 Styles and Sizes

No. 5 H. Selle Front Platform.

No. 42 H. Selle Giant Trussed Gear.
The largest and most successful builders buy their gears for the same reason that they buy their springs, axles and wheels.

No. 304. Selle Three Spring Express Gear.
We have the largest and most complete plant in the U. S. for the manufacture of wagon and truck gears. If we do not show what you want, can make it for you.

No. 72 H. Selle Truck Gear.
Dealers who do not now carry our line, should write for terms at once.

No. 17 B. H. Selle Wagon Gear.

No. 282. Selle Break Gear.

No. 61 H. Ready to receive the body.

No. 276. Selle Ambulance Gear.

THE AKRON SELLE COMPANY
AKRON, O. U S A

☞ WRITE FOR CATALOGUE No. 3

Please mention "The Blacksmith and Wheelwright" when writing to Advertisers.

CHOOSE YOUR GEAR. 1,000 styles and sizes of wagon gears were offered by this company. The smaller manufacturers could build bodies to special order of their customers, then mount them on an undercarriage that would suit the load to be carried.

147

ROAD WAGONS AND ROAD CARTS

THIS ROAD WAGON AT $25.95 IS A WONDER OF VALUE

while it is built in our own factory, covered by our one year binding guarantee and is much better value than you could possibly secure elsewhere, and is a grade of rig that is sold by other concerns at from $5.00 to $10.00 in advance of our price, we advise ordering one of our higher grade runabouts, one of the 1908 models shown on the following pages. Our profit is the same on this runabout as on our higher grades; therefore, to secure the greatest possible return for the money invested, we always urge our customers to select one of our higher grade runabouts, where you will get better wheels, better axles, better springs, stronger and better made body, better upholstering, a better finish and a more serviceable runabout.

$25.95

No. 11K900

DESCRIPTION OF No. 11K900

SEAT—29½ inches; skeleton back; tufted and padded cushion and back; lined seat ends; imitation leather upholstering. **BODY**—23 inches wide by 55 inches long; Corning style, well made, fitted with short carpet and Evans' enameled black duck dash. **GEAR**—Drop axles, 15-16 inch, double collar; full length axle caps; double reach, ironed and braced full bearing fifth wheel; three and four-plate elliptic springs; center bearing body loops. **WHEELS**—Sarven's patent style; 38 inches front and 42 inches rear; ⅞-inch rims, fitted with ¼-inch oval edge steel tires. **SHAFTS**—Hickory shafts, double braced, neatly trimmed. **PAINTING**—Body and seat black, neatly striped; gear, wheels and shafts a dark Brewester green, neatly striped. **TRACK**—4 feet 8 inches narrow or 5 feet 2 inches wide. Shipped from Evansville Indiana. Crated under 30 inches, weight 420 pounds.

No. 11K900 Price complete with shafts and steel tires....... **$25.95**

$32.15

No. 11K1015

TAKE YOUR CHOICE OF THESE TWO ROAD WAGONS

COVERED BY OUR ONE YEAR BINDING GUARANTEE

THESE ROAD WAGONS are better value than can be secured elsewhere at within $5.00 to $10.00 of our price, but our Blue Ribbon Runabouts for 1908 shown on page 95 are much better runabouts and are guaranteed for three years.

$34.15

No. 11K819

DESCRIPTION OF No. 11K1015

SEAT—Stick seat, made with bent sticks; seat handles; measures 29½ inches; 10-inch curved panel back; spring cushion; whipcord upholstering. **BODY**—23 inches wide by 55 inches long; piano box style; body is glued, screwed and plugged; short carpet; end panel lined with carpet; drill fiber boot on rear of body; leather dash; storm apron. **GEAR**—15-16-inch axles, dust and mud-proof bell collars, long distance spindles; three and four-plate elliptic springs; new style center bearing body loops; double reach, ironed full length. **WHEELS**—38 inches front and 42 inches rear; Sarven's patent style; ⅞-inch screwed rims, fitted with full ¼-inch oval edge steel tires. **SHAFTS**—Double braced hickory shafts, trimmed with leather 22 inches back from the point, flat straps; quick shifting shaft couplers. **PAINTING**—Body and seat painted plain black; gear, wheels and shafts painted blood carmine. **TRACK**—4 feet 8 inches narrow or 5 feet 2 inches wide. Shipped from Evansville, Ind. Weight, crated, under 30 inches, 440 pounds.

No. 11K1015 Price, complete with double braced shafts and steel tires.................................... **$32.15**

DESCRIPTION OF No. 11K819

SEAT—29½ inches; Georgia drop back; spring cushion, well padded and tufted; seat ends padded and lined; double bar nickel arm rails; upholstered in keratol leather. **BODY**—Piano box style, 23 inches wide by 55 inches long; swell body panels and convex seat panels; boot on rear of body; carpet; end panel lined with carpet; side panels finished inside; leather dash with nickel plated dash rail; storm apron. **GEAR**—$\frac{15}{16}$-inch axles; dust and mudproof bell collar long distance spindles; hickory wood caps; double reach, ironed and braced; three and four-plate elliptic springs; new style center bearing body loops; full bearing fifth wheel. **WHEELS**—⅞-inch screwed rims, fitted with ¼-inch oval edge steel tires; Sarven's patent style; 38 inches front and 42 inches rear. **SHAFTS**—Hickory shafts, trimmed with leather 22 inches from the tip, flat straps; double braced; quick shifting shaft couplers. **PAINTING**—Body and seat painted plain black, with neat design on seat riser; gear, wheels and shafts blood carmine, striped with black. **TRACK**—4 feet 8 inches or 5 feet 2 inches. Shipped from Evansville, Indiana. Weight crated under 30 inches, 435 pounds.

No. 11K819 Price, complete with double braced shafts and steel tires.................................... **$34.15**

EXTRAS ON ANY OF THE ABOVE ROAD WAGONS

Pole in place of shafts	$2.00
Both pole and shafts	4.65
Extra grade genuine leather upholstering in place of regular	2.00

$11.95

COVERED BY OUR ONE YEAR BINDING GUARANTEE.

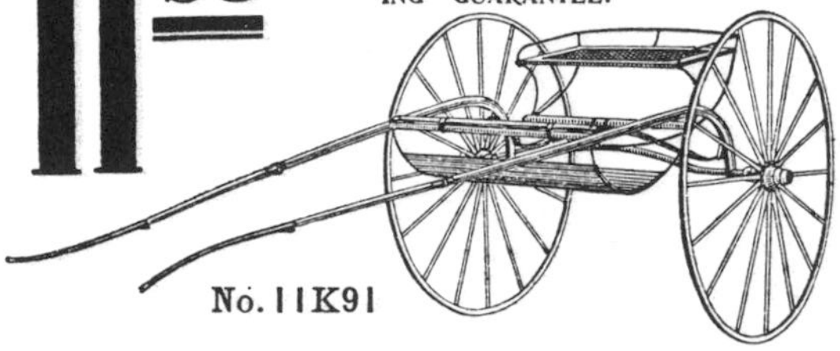

No. 11K91

IN BUILDING THESE CARTS

we use Sarven's patent wheels, with ⅞-inch rims; rims are fitted with oval edge steel tires; heavy hardwood spring block; seat frame, foot rack, etc., made of hardwood; hickory shafts; 1-inch double collar steel axle, square at the shoulders and round in the center with a high arch; easy riding oil tempered springs, hung so as to balance the seat perfectly. On our No. 11K91 we furnish a skeleton seat, with a seat rail, as shown in illustration. On the phaeton body cart, No. 11K101, we furnish a seat and lazy back, as shown in illustration. The seat is upholstered on the phaeton body cart in imitation leather. The body of the No. 11K101 is so constructed that the seat can be raised on hinges and small articles carried in the box under the seat. Both carts are built to carry two passengers, although the adjustment of the springs is so that they ride very easy with only one passenger. We use the same care in painting and finishing these road carts as we do our regular buggies. We paint the No. 11K91 road cart a rich carmine color. The body of the No. 11K101 road cart is painted black, with a rich blood carmine gear, neatly striped. We can furnish either one of these carts in the 4 feet 8 inches narrow or the 5 feet 2 inches wide track. Shipped from factory in Northern Illinois. Shipping weight, No. 11K91, 150 pounds; No. 11K101, 180 pounds.

COVERED BY OUR ONE YEAR BINDING GUARANTEE.

$14.65

No. 11K101

No. 11K91 Skeleton Cart, as shown in the illustration. Price, complete.................................... **$11.95**

No. 11K101 Phaeton Body Cart, as shown in the illustration. Price, complete.................................... **14.65**

WAGONS AND BUGGIES. Two pages from the 1908 Sears, Roebuck & Co. mail order catalog.

Credit: Author's Collection

AUTOMOBILE SEAT TOP BUGGY — WITH WING DASH — $59.95

MADE THROUGHOUT OF HIGH GRADE MATERIAL, A VERY POPULAR STYLE OF OUR SOLID COMFORT LINE; REMOVABLE TOP FITTED ON A SHIFTING RAIL WITH PATENT FASTENERS; TOP CAN BE REMOVED SO THE JOB CAN BE USED AS A RUNABOUT; A HIGH CLASS STYLISH UP TO DATE AUTOMOBILE SEAT TOP BUGGY WITH THE LATEST STYLE WING DASH.

DESCRIPTION

SEAT. A very handsome design of automobile seat, roomy, comfortable, with a high bent back fitted with plenty of soft resilient coil springs, upholstered and tufted as shown in the illustration, including the seat ends; cushion is of the box frame pattern, neatly tufted and upholstered and fitted with plenty of soft coil springs; the seat measures 29½ inches across top of cushion. We use either an all wool dark green (nearly black) imported body cloth or a soft high grade leather in upholstering this seat. If any choice, be sure to state in order, otherwise we will use our best judgment, according to the section of country to which the job is to be shipped.

BODY. Well made, strong substantial body, 23 inches wide by 55 inches long; swell panels, made of the best seasoned material; hardwood frame, sills, corner posts and seat frame; well made in every particular, will outwear two ordinary bodies. Boot covering body rear of seat; full length velvet carpet; special design automobile bent front wing dash, made of high grade patent leather, padded.

TOP. Full sweep graceful three-bow genuine leather quarter top, with the quarters cut deep, of extra high grade enameled top leather; wide genuine leather back stays; heavy rubber roof and back curtain; heavy raised valance, stitched front and rear. The top throughout, including back curtain, is lined with a very substantial heavy dark green wool head lining to match the upholstering in the seat; four roll up straps; heavy waterproof rubber side curtains and storm apron. If wanted, we will furnish a four-bow top instead of a three-bow top, without extra charge.

GEAR. Special design graceful arch axle, fitted full length with a hickory wood axle cap, cemented, sanded and clipped to the axle; spindles are of the long distance dust and mudproof pattern, 15-16-inch bell collar; double reach, ironed full length, stayed and braced with wrought iron stays; large 11-inch full circle easy turned fifth wheel; 36-inch open head, elliptic, oil tempered easy riding springs, attached to body with our celebrated center bearing wrought iron body loops; steel plates running full length on body from one body loop to the other; a strong, substantial gear throughout, one that will outlast two ordinary factory grade gears.

No. 11K628

$59.95

WHEELS. Extra high grade Sarven's patent wheel with selected second growth hickory spokes, sixteen to each wheel; bent hickory rims, ⅞ inch wide, fitted with a full ¼-inch oval edge steel tire, bolted between each spoke. Each wheel is carefully tested and tried, guaranteed to run true, have the proper set, proper gather and the proper dish; front wheels 38 inches high, rear wheels 42 inches high. Can furnish the Columbus staggered spoke wheel if ordered at $1.50 extra. (See page 92). Can furnish wheels 40 inches front and 44 inches rear, if desired, without extra charge.

SHAFTS. Selected grade XXX shafts, neatly trimmed with 30-inch shaft leathers, round straps and double braced, also fitted with a celebrated Bradley quick shifting shaft coupler.

PAINTING. This Automobile Seat Top Buggy is given a first class job of painting and finishing, the body and seat are given a mirror finish and a jet black color, no striping; gear is painted a rich dark blood carmine, mirror finish, neatly striped with black. The whole appearance of the job with a plain black body and a rich blood carmine gear, neatly striped, makes a handsome appearance; it has just enough life without being at all flashy or loud. Can furnish dark Brewster green gear instead of blood carmine if ordered.

TRACK. 4 feet 8 inches or 5 feet 2 inches. State width wanted.

No. 11K628 Price, complete with double braced shafts and steel tires $59.95

Price, complete with ¾-inch Kelly Springfield guaranteed rubber tires.. 73.05

Price, complete with ⅞-inch Kelly Springfield guaranteed rubber tires. 74.90

Price, complete with 1-inch Kelly Springfield guaranteed rubber tires. 77.80

(⅞-inch rubber tires are the best suited for this top buggy.)

Covered by our regular binding three-year guarantee to be free from defect in either material or workmanship.

EXTRAS.

Pole in place of shafts with Bradley couplings $2.00
Both pole and shafts with Bradley couplings 4.90

Shipping weight, crated under 30 inches, about 475 pounds. Shipped from Evansville, Indiana.

HEAVY CONCORD TOP BUGGY

TWO-YEAR GUARANTEE.

DESCRIPTION OF No. 11K263

$61.45

SEAT—Special style seat, large and roomy; solid panel spring back and spring cushion; upholstered in genuine leather. **BODY**—28 inches wide by 58 inches long; panels, 8½ inches; hardwood frame; lined panels; strap leather boot on rear of body; leather dash. **TOP**—Leather quarters and leather back stays; back stays padded and lined; lined back curtain; rubber side curtains; wool head lining; rubber roof; full corded top; storm apron; four bows. **GEAR**—1 1-16-inch axles; dust and mudproof long distance spindles; hickory axle caps; full Concord side springs, hung on equalizers, attached to body with hickory spring bar; three perch gear; full bearing fifth wheel. **WHEELS**—1-inch oval edge steel tires, bolted between each spoke; Sarven's patent style; 38 inches front and 42 inches rear. **SHAFTS**—Double braced hickory shafts, neatly trimmed with leather; quick shifting shaft couplers. **PAINTING**—Body, plain black; gear, wheels and shafts a dark Brewster green, neatly striped. **TRACK**—4 feet 8 inches or 5 feet 2 inches. State width wanted. Shipped from factory. Weight, crated under 30 inches, 600 pounds.

No. 11K263

No. 11K263 Price, complete with double braced shafts and steel tires $61.45
No. 11K2635 Same job, only hung on drop axle and elliptic springs instead of Concord 60.15

EXTRAS.

Pole in place of shafts $2.25
Both pole and shafts 5.00
Full leather top with rubber side curtains in place of leather quarter top 4.00
Brake 5.00
1⅛-inch wheels 2.30

1908 MODEL PHAETON

GUARANTEED TWO YEARS.

DESCRIPTION OF No. 11K409

$61.95

SEAT—Special style; panel spring back and box spring cushion; seat panels lined; seat measures 32 inches across top of cushion and 19 inches deep; upholstered in either a heavy dark green body cloth or extra grade upholstering leather. **BODY**—Phaeton style, very roomy; hardwood frame; oil burning lamps; fenders; carpet; leather dash. **TOP**—Leather quarter top; rubber roof and back curtain, lined raised valance, stitched front and rear full corded top; rubber side curtains; patent curtain fasteners; leather back stays, padded and lined, storm apron; three-bow top. **GEAR**—1⅛-inch axles, double collar; hickory axle caps; double reaches, ironed full length; full bearing fifth wheel; three and four-plate oil tempered elliptic springs. **WHEELS**—Sarven's patent style; 36 inches front and 44 inches rear; ⅞-inch screwed rims, fitted with full ¼-inch oval edge steel tires, full bolted. **SHAFTS**—Hickory shafts, nicely trimmed with leather 34 inches from the point; double braced; quick shifting shaft couplers. **PAINTING**—Body panels black, neatly decorated, Brewster green belt around edge; gear, Brewster green, neatly striped. **TRACK**—4 feet 8 inches or 5 feet 2 inches. State width wanted. Shipped from factory. Weight, crated under 30 inches, 500 pounds.

No. 11K409

No. 11K409 Price, complete with double braced shafts and steel tires $61.95

EXTRAS.

Pole in place of shafts $2.25
Both pole and shafts 5.00
⅞-inch Kelly Springfield guaranteed rubber tires 14.95

CHAPTER IX

THE HORSES

Ever since the sixth century B.C., when the Scythians ruled the regions north of the Black Sea and mastered the horse, it seems any race of people who conquered parts of the world were aware of the value of the horse.

Not only were they aware of the horses's value in overwhelming and conquering in military escapades, but they realized its value for domestic purposes.

Once domesticated, horses proved to be the world's most versatile work animal. They provided the power and mobility that enabled man to forever move forward.

The Right Horse for the Job

The importance of the horse to man grew as the centuries rolled by. New breeds were developed to handle specific tasks.

Horses used for heavy draft work, pulling carriages, military purposes and riding gradually developed certain characteristics through selective breeding.

The varied sizes, uses and values of horses were illustrated by reports in prime horse markets. The July 6, 1892, *Breeder's Gazette* quoted horse prices in the Chicago market:

"Streeters—$95.00 to $115.00

Chunks—$100.00 to $135.00; 1,050 lbs. to 1,200 lbs.

Express horses—$160.00 to $215.00

Heavy Drafters—$160.00 to $240.00; 1,500 lbs. to 1,700 lbs.

These are five to eight years old and in

MAMBRINO-BACCHUS. The Mambrinos were an important family of trotting horses. This coach horse was developed by crossing a Mambrino with a Thoroughbred.

HIGHCLIFFE. Champion Cleveland Bay at the 1892 Columbian Exposition.

PALADIN. French Coach horse (C-1900).

150

THESE TWO German Coach horses were Champions in 1893.

CLEVELAND BAYS. These were considered a general utility horse, used for light farm work, harnessed to a buggy or a carriage. Bay with black legs, they usually stood about sixteen hands or a little better.

PERFECTION. Champion French Coach Stallion.

good flesh, well broken and sound.''

Also listed were coach horses, road horses, a carriage team, gentleman driver's span, cob horse, combination saddle and harness horse. The article stated, ''good shapy chunks and toppy drivers sold very well, common green horses sold with difficulty.''

A year later, the same magazine reported the following prices for July for the Chicago market:

Bay Road Horse	1,050 lbs.	$175.00
Chestnut Road Horse	1,000 lbs.	$192.00
Brown Road Horse	1,000 lbs.	$225.00
Bay Coach Horse	1,000 lbs.	$190.00
Gray Coach Horse	1,250 lbs.	$170.00
Gray Express Horse	1,500 lbs.	$170.00
Bay Draft Horse	1,700 lbs.	$220.00
Gray Draft Horse	1,500 lbs.	$155.00

Prices were higher in 1914 when the Breeder's Gazette reported, ''Heavy expresser types, upstanding, high-geared geldings weighing a little over 1,500 pounds sold for $285.00. One 1,700-pound gelding sold for $300.00, and a group of five 1,400-pound horses sold for $245.00 each.

''Good wagon horses sold for $180.00 to $200.00 in 1917, and in the Boston market 1,600-lb. horses were bringing $275.00 to $300.00.''

For a time, there seemed to be a lot of fuss about developing a new breed of horse. In the July 1, 1915, Breeder's Gazette, an article said, ''We need a new breed of horse 1,200 to 1,300 lbs. for buggy and draft work. Chunks and Gunners (artillery horses) wide at both ends, deep in the middle, going straight and true are needed—15.1 to 16 hands and weighing 1,300 lbs.''

In the issue two weeks later, July 15, the Gazette quoted an article from the Cornbelt Farmer Magazine, ''Try as we may to breed them either fast or big, the general purpose horse will not down. The country is full of in-between sorts. Some are pushing for a new Breed so as to get Gunners with regularity and

THE HORSES

certainty. The history of horse breeding reveals the error of this view.''

Draft Horses

DRAFT HORSE FACT: An interesting misconception about draft horses is that they pull loads; as a matter of fact, they push loads. As horses push against their collars, their strength and power are converted to the pull on the traces, thus to the singletrees.

There are five major breeds of draft horses in America:

Percheron: (From a 1936 booklet, ''How to Select Percherons'' by Ellis McFarland, Secretary of the Percheron Horse Association of America.) ''It is of French origin. They are black or grey. The horse should be thick set, short backed, deep bodied, heavy boned, well muscled, good going, medium size, nice head and neck and plenty of quality. Mares should be 16 to 17 hands and stallions 16-1/2 to 17 hands. Stallions should weight 2,000 to 2,200 lbs.''

Suffolk Punch: (Excerpts from booklet published by American Suffolk Horse Association in 1938) ''It is the most docile of draft breeds, fastest walkers, easiest keepers, solid chestnut in color, even tempered, easy going, extreme endurance, English origin, whole appearance of the Suffolk is pleasing rotundity without flatness or grossness anywhere, stands average 16 hands and weighs 1500 pounds to 1600 pounds.''

Belgian: (From booklet published by Belgian Draft Horse Association, 1938)

''The origin is Belgium. Ancestors were large black Flemish horses. Predominate color today is chestnut, roan and bay. The horse had a good finish, style in head and neck,

HIGH STEPPING Hackneys were used on coaches and carriages. This stallion was a prize winner in 1898.

THE MORGAN horse was a great favorite for carriages and buggies. This breed was stylish in appearance, had a fine head and was tractable, but with a nice spirit. Photo was taken around 1905 near Newbury, Vermont.

PRIZE WINNERS. Ad in Dec. 20, 1893, Breeder's Gazette for coach horses.

Credit: Author's Collection

Credit: Vermont Historical Society

Credit: Author's Collection

CHOOSE YOUR BREED. In 1892 horse dealers had a large selection of breeds in ads in the Breeder's Gazette. Offered for the carriages, buggies, and coaches were Cleveland Bays, Hackneys, German Coach Horses, Mambrinos, and French Coach Horses.

THE HORSES

sloping shoulders and pasterns, strong level back and shapely feet."

Shire: (From booklet, "The Sovereign Shire", by J.G. Truman, President of the American Shire Horse Association in 1913.) "English origin. The Shire is the bulkiest draft horse, heaviest boned and inch for inch the most powerful horse."

Clydesdale: (Excerpt from book on various breeds by R.S. Summerhays.) "Developed in Scotland. Average height 17 to 18-1/2 hands, average weight 2,000 pounds. Outstanding characteristic are combination of weight, size, and activity and wearing qualities of feet. Head broad across eyes, front of head neither dished or roman. Well arched and long neck, short back and well sprung ribs. Color bay, brown or black and much white on face and legs. Active mover for its size."

In 1920 the U.S. Department of Agriculture reported that there were 70,613 purebred Percherons, 10,838 purebred Belgians, 5,580 purebred Shires, and 4,221 purebred Clydesdales in the country.

"Market for Draft Horse Type," said an article in Sept. 1, 1909, *Breeder's Gazette,* "Buyers demand a wide deep chest, a strong short back, close coupling, and full deep barrel. These indicate body vigor.

"Especial importance is attached to large hooves under big boned, flat, smooth legs and strong clean hocks. Clean straight face, wide forehead, large clear eyes.

"A horse like this weighing 1,725 pounds, 16 hands will bring $365.00 to $415.00. Contrast these market toppers with the rough, unshapely, leggy, cherry picker that has a nar-

Buying a Horse.

Some one who has evidently "been there and knows a thing or two," says: "Don't buy a horse in harness. Unhitch him and take everything off but his halter and lead him around. If he has a corn or is stiff or has any other failing you can see it. Let him go by himself a ways, and if he staves right into anything you know he is blind. No matter how clear and bright his eyes are he can't see any more than a bat. Back him, too. Some horses show their weakness or tricks in that way when they don't in any other. But be as smart as you can you'll get caught sometimes. Even an expert gets stuck. A horse may look ever so nice and go a great pace and yet have fits. There isn't a man who could tell it until something happens. Or he may have a weak back. Give him the whip and off he goes for a mile or two, then all of a sudden he stops on the road. After a rest he starts again, but he soon stops for good and nothing but a derrick can start him.

"The weak points of a horse can be better discovered while standing than while moving. If he is sound he will stand firmly and squarely on his limbs without moving them, with legs plump and naturally poised; or if the foot is taken from the ground and the weight taken from it disease may be suspected, or at least tenderness, which is the precursor of disease. If the horse stands with his feet spread apart, or straddles with his hind legs, there is a weakness in his loins and the kidneys are disordered. Heavy pulling bends the knees. Bluish, milky-cast eyes in horses indicate moon-blindness or something else. A bad-tempered one keeps his ears thrown back; a kicking horse is apt to have scarred legs; a stumbling horse has blemished knees. Then the skin is rough and harsh and does not move easily to the touch; the horse is a heavy eater and digestion is bad. Never buy a horse whose breathing organs are at all impaired. Place your ear at the heart and if a wheezing sound is heard it is an indication of trouble."

WORDS OF WISDOM. "Buying a Horse" appeared in the Oct. 18, 1893 Breeder's Gazette.

THE PERCHERON horse predominated on city streets and the farm the last 30 years of the horse-drawn era. This dapple grey stallion, "Seducteur", is a splendid example of the breed.

HEADING EAST. This photo shows a group of 20 Percherons in Chicago ready to be shipped to New York and Jersey City by Swift & Co. These horses are five to seven years old and weigh 1350 to 1800 lbs. They cost an average of $200.00 per head.

LAET 133886, International Grand Champion Stallion, 1921; Ohio State Fair Grand Champion Stallion, 1919

LAET — THE SIRE *of* SIRES

A BREEDING RECORD UNEXCELLED   LAET BLOOD BREEDS ON

For Sale!

One 3 yr. old imported dark grey stallion, one 2 yr. old imported black stallion, both sound, drafty, good action and of true Percheron type; also one 2 yr. old imported dark grey filly.

WOODSIDE FARMS

W. H. BUTLER   R-2   *Write to* DAVE HAXTON
Owner   Columbus, Ohio   *Manager*

(*Mention Percheron Review*)

Credit: Author's Collection

UNEXCELLED. ''Laet'' was the outstanding Percheron in the 1920s and 1930s. ''The Sire of Sires'', said owner, W.H. Butler.

row chest, is goose rumped, has pinched shoulders, a characterless head, is slabsided —it will bring $175.00.''

Coach Horses

The horses used to pull privately owned vehicles were of special breeds and lighter weight.

There were French Coach horses, German Coach horses, Hackneys, Morgans, Cleveland Bays and Oldenburgs. These horses showed good action, which was an absolute requirement.

These coachers with a trappy gait could move rigs at eight to 10 miles per hour.

A fancy team of two or four coach horses was a status symbol in the period 1870 to about 1900. These horses pulled the Broughams, Victorias, Landaus, Road Coaches, and Park Drags. Their tails were bobbed and crimped and heads were held unnaturally high by checkreins.

The Coach horses averaged 16 hands and weighed 1,100 to 1,250 pounds.

For everyday buggy work on the farm, or

THE HORSES

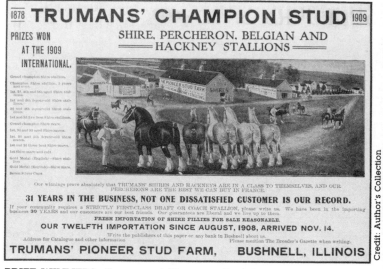

PRIZE WINNERS. Trumans' Pioneer Stud Farm said in this 1910 ad, "31 years in the business, not one dissatisfied customer is our record".

for people living in smaller towns, and some in the cities, the horses used were grades. Steady, quiet, easy-going horses were needed because frequently the women would drive to church picnics and socials, or go shopping. Very often gentle aged mares were used.

Buying, Selling and Trading Horses

Dealers, buyers, and breeders of horses had a whole range of expressions that were special to the business.

Arnold Hexom, Iowa Percheron breeder and horse auctioneer, recalled many old-time phrases used by horse dealers when selling. For example, "This team is tougher than a cooked owl," or, "Now there is a real butter and egg horse," meaning it was a dependable single driving horse. A "clobber" was a horse with an awkward gait. Then there was a "pussy footer," a short strider that had a graceful gait. He said a good horse would be sold as "well broke, level headed, honest work horse with a heart as big as a barrel." "Of course," Hexom added, "there were always a few horses that were a little bronky, or were snorters, or cold shouldered. Then there were some that were pigeon toed, pig eyed, droop eared or a cake walker."

"Picking good horses in the first place pays off", said James G. Boyd, who, for 17 years, has selected the horses for Pabst Brewery Co. Under his supervision at Milwaukee there are 175 head. Scattered around at other plants there are several hundred more, making a total of 700 horses for which he is responsible.

"He buys only horses that have good sound feet.

"Last year he only sold 25 head out of 700

PERCHERONS were featured on this front cover.

156

THE BREEDER'S GAZETTE

A WEEKLY JOURNAL OF
LIVE STOCK HUSBANDRY.

VOL. XXIV.   CHICAGO, ILL., SEPTEMBER 6, 1893.   No. 10.—615.

FOUR-YEAR-OLD STALLION PRINCE PATRICK 6773 (8933)—PROPERTY OF N. P. CLARKE, ST. CLOUD, MINN.
FIRST-PRIZE FOUR-YEAR-OLD AND CHAMPION CLYDESDALE STALLION ANY AGE AT WORLD'S COLUMBIAN EXPOSITION.

PRINCE PATRICK. Clydesdale stallion was featured on this front cover of the Breeder's Gazette.

THE HORSES

for replacement." (*Breeder's Gazette* Aug. 21, 1912)

Horse traders were a breed unto themselves.

The "road traders", or "roaders", were itinerants who moved through the countryside in horsedrawn covered wagons. Tied behind their wagons was a string of plugs and snides with, perhaps, one or two good horses mixed in. They would trade in horses or mules.

"Barn traders", operated at a livery stable. Perhaps the trader owned it, or maybe operated out of it with the owner's permission. He was apt to be more straightforward in his dealings, since he was more or less a permanent fixture in town.

The "roaders" were here today and gone tomorrow, and perhaps, covered the same territory only once every two or three years. Their philosophy bordered on outright "trimmin', skunkin' and skinnin' " a man. They were adept at hiding, temporarily, all the defects of a horse. Rarely did these "roaders" trade even up one horse for another. In most of the transactions, there was boot money, particularly if the "roader" sensed the customer really desired the horse in question.

If a customer discovered a defect after a deal was cut, the "trader" would very often back trade, trading back horse for horse. Somehow the "trader" always kept the boot money, or products that were in on the original deal.

"Roaders" loved to dicker. It seemed on any trade they ended up with a horse that they, in turn, could trade along the road for even better merchandise.

They were an unscrupulous lot, but then

Credit: Author's Collection

WINNING SIRE. "Carnot" was a super Percheron stallion. Owner W. S. Corsa listed "Carnot's" winnings in this 1912 ad to impress buyers who might be interested in his progeny.

they simply figured they were matching wits with a man. Whoever lost could chalk it up to experience.

Horses and mules afflicted with quite an array of ailments could be windbroken, foundered, galled, spavined, sweenied or gravelled. To treat what ailed them, there were endless assortments of bottles and cans and tubes of various liniments, laxatives, ointments and leg washes. There were healing ointments for cuts, wounds, scratches and cracked heels and blistering ointments for spavins, ring bone and splints. One product— a most wonderful oil—sold for $1.00 a bottle and was good for contracted muscles, thick-

Horses for Heavy City Work.

An old subscriber to THE GAZETTE, Mr. C. A. Campbell, who is an extensive retail coal dealer in Boston, has contributed to the *Coal Trade Journal* an article on the "Suitable Horse for the Retail Coal Dealer," a copy of which he has kindly furnished us. And inasmuch as the points brought out by Mr. Campbell are equally worthy of consideration by horse-breeders we take the liberty of making a few extracts:

Formerly in the city the average load of a two-horse string team, attached to a dead axle, two-wheeled cart (there were no others in use), was 4,000 to 4,500 lbs.; now, with a breast team on a four-wheeled, slanting bottom, spring wagon, a load of four to five tons is easier and quicker handled. The horses should be from 15.3 to 17 hands high, from 1,400 to 1,900 lbs. each in weight, and generally matched in color, but of necessity alike in other respects, as to handle loads of this size they must work together. In selecting horses for this work the following general rules should apply: The ears short and pointing forward, forehead broad, eye prominent and clear, lips firm; neck inclined to be short, deep where it joins the body; well muscled, withers in line with neck, not too high or sharp; shoulders moderately straight and well muscled, well ribbed up; breast broad and full, barrel round, back short, loins broad, croup full, not too sloping; tail well carried, fore-arm large and well-muscled, knee of good size and strong, front cannon short and flat, pasterns short and strong, fore-feet full, round, not flat; frog well developed and elastic, heels full and uniform, upper thigh heavily muscled, thick and broad; gaskin strongly muscled, hock large and strong, hind cannon broad and flat, pasterns short and strong, hind feet smaller and more concave than front ones, temperament kind, but having energy; not nervous, in action a good walker, with some style. Color, never mind; a good horse is never a bad color.

For handling loads in crowded or narrow streets, or in out-of-the-way places, a one-horse iron cart with five-feet wheels and springs is now much in use. The suitable horse for this work is rather difficult to find, as weight and first-class conformation, with less height and more breadth are required. The breeding of "chunks" has been neglected, as their use has been limited, and they are not liked on the farm; yet good ones can be found, and while they are usually moderate, yet they are powerful, generally long-lived and very profitable. My father purchased such a one at four years of age that worked almost daily for twenty-four years until last year, when on account of an accident she was killed. The object of low wheels on a cart is that the line of draft is brought in bearing with the slope of the shoulder; they also cause less shock from the weight of the load thrown on the horse's back, relieving the muscles of the legs, the tendons and the feet. A cart at best is a hard place to put a horse, and I wish I could impress on every dealer the absolute necessity of keeping the collar and saddle in good condition in order to get the best results, also to see that the breeching is not too low, as it is in most cases; have it well up on the haunches; the horse can back better than when it is lower down. This is equally true of horses in wagons.

For long distances, dirt roads, hills, and bad places generally (and outside of large cities they are the rule), a "general-purpose" horse is wanted that is yet to be bred. The mating of the draft and coach, or the draft and trotting horse, has not succeeded in producing a type, the product resembling one parent more than another, while the second generation is as fitful as the first, generally declining to resemble much of anything. The horse for this use should be lighter, weighing from 1,200 to 1,400 lbs., more rangy and generally more symmetrical than the heavy draft horse, say an exaggerated coach horse, similar to the horses of the Adams and American Express Companies. For this class there is always a great demand, and the dealer who has a team that can fill this place is fortunate. The wagons and harnesses should be much lighter and smaller than the ones for slower work, carrying about three tons, and always with springs, which are a great relief on a dirt road.

In a large establishment the animals for the single wagons are not so hard to obtain, most dealers furnishing enough from their heavier horses that have outgrown their usefulness in their original places, and that if put to lighter work have many years of earning capacity before them. Horses for such work are always plenty and cheap. No doubt I will be laughed at for being so sentimental as to never sell or trade a horse which, when I have fairly tried, is what I want, but such is my rule, and I find places where old and young alike are useful as long as they are well and can eat their rations; when they cannot they are mercifully disposed of.

Having secured such a horse as is desired it is taken to the stable, the age, height, weight, color and marks recorded, and right here the trouble begins. No doubt the animal purchased is perfectly broken, but to what? Entirely different work and uses from my requirements; therefore I find it far easier and producing better results to re-educate gradually; by so doing the horse is getting acclimated, growing stronger, getting way-wise, the muscles hardened, and in from two to four months is ready for years of hard and constant labor. The food of the green horse is of vital importance, as from having been brought up on bulky food it is now necessary to feed grains and muscle-making material, the change being slow and gradual. Let the first meal be a warm mash well salted, afterward feed twice daily about three quarts of oats with a grain ration at each meal, with mash twice a week for a few weeks, till the better condition and regular work of the animal will keep him in good condition with a mash every other Saturday night. I have found this feeding to allay the sickness which almost always comes to the green horse and prepares him for the heartier feeding and harder work of the future.

The work at first should be moderate, not over half a day at a time, with light loads the first two or three months, by which time, if healthy and in good condition, the animal is ready to fill the place selected.

Among the very first and most important matters to be attended to is the shoeing, about which very few dealers have any knowledge, everything being left to the farrier, who frequently knows less than the dealer. The foot of the horse from the country is never in condition for hard work and will require several shoeings before it can be brought to the proper shape, which, once done, and properly shod as often as needed—surely once in four weeks—can be kept in good condition on a regularly worked horse as long as the body lasts. The foot of the horse has always been to me a subject of great interest and study, the saying, "no foot, no horse," having been early impressed upon me. The secret of excellence in the horse is good, sound, well-balanced feet, without which the conformation becomes changed and we have an entirely different animal. The leveling, balancing, general preparation and care of the "horny box" containing the wonderful mechanism that when in harmony gives the horse the ability to accomplish the best results demand more time and space than can be given in an ordinary article. From careful observation I feel safe in saying that more than half of the lame and decrepit horses can be cured by proper attention to and care of the feet. My new and lame horses have my personal attention when at the forge, and I almost invariably have the pleasure of being successful. It takes time, and to some would be disagreeable, but it pays.

The feeding should be regular, and the grain and hay of best quality. In my stables the young horses get three to four quarts of the best oats in the market three times a day, with twelve to fifteen pounds of the best hay, most of it at night, while the older horses get a pint of whole corn with their oats. With this ration the horses are kept in good condition, not fat, and are generally in good health.

CITY WORKERS. "Horses for Heavy City Work" was published in the Oct. 18, 1893, issue of the Breeder's Gazette.

THE HORSES

ened tendons, shortened gait, spavins, founder, strains, rheumatism, growths, boils and calluses.

Still another remedy claimed it could cure wind puff, capped hock, fatty tumors, puffs and swellings. Other products eradicated worms.

A runaway horse, particularly on city streets, was a frightening experience for those in the vehicle and those driving other carriages or buggies. When an unpredictable horse spooked at blowing paper, or an unexpected explosion of some sort, it galloped with utter abandon down the street headed for catastrophe unless the driver could get the animal calmed down and back under his control.

Inexperienced horsemen, or those who were utterly incompetent, had most runaways. There were some horses who were inclined to be wild eyed and easily excited. A good horseman would sell such an animal, but if kept and used he would be ever alert and through patience and quiet approach bring the horse around. The wrong man, shouting and cursing and yanking a horse around, could soon ruin a perfectly good animal.

Cruelty to horses was overlooked and ignored until a New York City citizen, Henry Bergh, stirred up the public to such an extent in 1869 that his movement for "Society for the Prevention of Cruelty to Animals" was accepted and swept the country. His efforts were concerned with the handling and treatment of all animals, but his main focus was on the everyday abuse of cart and carriage horses on the streets. Bergh was infuriated at indiscriminate flogging of horses struggling with loads too big for them to handle.

The U.S. Department of Agriculture, in their 1869 yearbook, stated, "Simple humanity would dictate that the services of useful animals should be rewarded by proper care and attention."

The basic problem was one of ignorance and also indifference on the part of some horse owners and drivers.

In the September 16, 1908 issue, the *Breeder's Gazette* touched on this point, "The horse is a machine capable of doing so much work if full fed, capable of doing less according to the manner of his feeding and the work he has to do.

"Underfeed him and overload him and you can soon tell how much he can do, or rather cannot do. When he is underfed and overloaded, the driver too often essays to mark up the difference in yelling—and vociferousness is the poorest sort of horse feed.

"The more quietly he is treated the better he will work.

"He is a poor horseman who seeks to hide his inefficiency behind a loud mouth. Whether the work be in the field, or on the city street the driver who uses his voice to make his horses do his work has much of horsemanship to learn."

The American Society for the Prevention of Cruelty to Animals in New York City had 12 horse ambulances that they operated. If a horse was severely injured it was shot. Most calls came in extremely hot weather, or during severe winter conditions such as icy streets. In 1916 alone over 50,000 horses died in New York City.

As street congestion increased in the big cities various companies began to haul larger and larger loads to reduce trips to the railroad yards, docks and warehouses. The less trips the less waiting for a spot to unload or pick up.

A story titled "Big Geldings" that appeared in the July 8, 1908 issue of the

The sextet in gray.

A big black six.

OLD! Page of photos taken at the New York Horse Market in 1914. The *Breeder's Gazette* ran the story. The photos show the quality of horses that were quickly purchased.

The Horses of Swift and Company.

TO THE GAZETTE:

In compliance with your request for information concerning the horses and vehicles in our stables we may say that we ordinarily keep at least 225 horses for use in our various vehicles in the stock-yards, about the packing house and in the city. In the yards we use about thirty "tip-carts" (single horse) for hauling offal, and also a number of "roustabout" wagons for handling rough tallow and the like and for general teaming. To carry our stenographers and lady clerks from the station through the yards to our offices we keep two park phaetons and one wagonette, the total carrying capacity of which is fifty people. For city teaming—freight-depot work and beef-delivery to our trade—we keep eighty-five wagons of four different sizes, what we call our standard sizes, and in addition three four-in-hand stake trucks and one six-in-hand box wagon. Our four standard sizes are as follows: An eight-foot single-horse wagon on which we load a ton; an eight-foot-six-inch single-horse wagon which carries 3,000 lbs.; a ten-foot two-horse wagon for 4,000-lb. loads, and a ten-foot-ten-inch two-horse wagon which is loaded with 6,000 lbs.—three tons. For the wagons which carry a ton we use horses of from 1,250 to 1,300 lbs.; on the wagons hauling 3,000 lbs. we put horses weighing 1,400 to 1,450 lbs.; the teams which draw the two-ton loads are from 1,200 to 1,250 lbs., active and light horses, while for the three-ton wagons our horses will average 1,500 lbs. These weights are for hardened horses. When we buy we generally allow for a shrinkage of 100 to 150 lbs. in hardening down to work. Our three wagons for four-in-hands are stake trucks weighing 4,300 lbs., which carry six tons of packed products of various kinds turned out from the packing plant. We use these four-in-hand trucks altogether in our freight-depot work. The six-in-hand (which you illustrate in this issue and which was first at the Columbian Exposition in the great contest described in your report of the cart-horse parade) we use exclusively for the delivery of beef and provisions to our Chicago trade. As will be seen we use a box wagon (we never haul beef in a stake truck, as it is impossible to keep it clean except in a box), which weighs 4,600 lbs., and we load it with eight tons, making a total weight, wagon and load, of ten and one-third tons. Such in brief is a description of the character of our vehicles, the weights of our horses and our ways of loading.

Concerning the horses themselves we have already informed you that we buy only grays and roans. The latter are not plentiful so gray may be said to be the prevailing color in our stables. We have scarcely twenty of any other color. The fact that we use exclusively Percheron horses which are mostly grays led us to adopt this color as a sort of a trade mark. We buy all our horses on the Chicago market. Mr. J. M. Shaw, the superintendent of our stables, who has had ten years' experience in this line, gives his personal attention to the selection and matching of our horses and is at all times on the lookout for the kind we use. All our branch establishments are horsed from this market. We are working probably 2,000 horses in cities from Denver to Boston and all are bought on the Chicago market and shipped as needed. We buy from 150 to 200 fresh horses a year; this comes about through general wear and tear and the extension of business. Last year we lost only four horses in our Chicago stables and two of them died of blood-poisoning occasioned by picking up nails in their feet, but of course granite blocks and railroad tracks in a city disable more or less horses in a year's time.

We have already expressed our preference for the Percheron. We believe this horse has the best foot, leg, action, disposition, and endurance, and we use no other breed even in our carry-alls. These horses are active, strong, resolute, tireless walkers, and have the heart to stand their work. We can get them of the weights we need. We have no use for thick-legged or hairy-legged horses. Their legs are always sore in spite of careful grooming.

Most of the horses in our stables come from McLean Co., Ill., about Bloomington and Normal, where, as your readers are aware, farmers have long been extensively engaged in breeding Percherons. Many of them have been bred by Messrs. Dillon and Stubblefield. We have also found some excellent horses on this market from Central Iowa but we do not know from what particular counties they come.

Our horses have a good ten hours' work a day. They are fed three times a day from six to eight quarts of oats at a feed. We buy nothing but No. 2 white oats. We use loose timothy hay which is always before the horses when in the stable and we bed with the best rye straw. The noon feeds are generally taken from nose-bags.

Concerning the six-in-hand which forms the subject of your illustration we may say that they are a cross-matched team. The off horses are chestnuts with light manes and tails. Each horse has a white strip in the face. This is the only mark on them excepting a very small white spot on the hip of each, which is not larger than a ten cent piece. They are six, seven and eight years old, and the wheel and lead horses are full brothers. We know nothing about their breeding except that they are Percherons. Mr. Shaw came across them one day in the yards and bought them from their owner who intended to take them to New York. The chestnut in the swing pair came from Iowa. The near-side horses are dapple grays with white manes and tails and are five, six and seven years old. They came from Bloomington. Taking them as a team they are very closely matched as to style, quality, height and weight. These horses we matched up ourselves and have driven them together about one year. They are very handsome and will attract attention anywhere. There is not a "pimple" on them.

SWIFT AND COMPANY.

THE EDITOR of the *Breeder's Gazette* wrote to Swift & Co. asking about their draft horses, the work they did, what loads they pulled, types of horses, and other questions. Their answer, fascinating and informative, was printed in the December 20, 1893 issue.

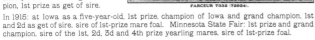

GRAND CHAMPION RETIRES. For three years Farceur had been an outstanding Belgian Stallion on the show circuit for horse dealer Wm. Crownover. In this ad in the December 16, 1915, issue of the *Breeders Gazette*, Mr. Crownover announces Farceur's retirement from the show ring.

THE HORSES

Breeder's Gazette explained how the horse industry tackled the problem of congestion on the streets:

"Advances in commercial circles had rendered the use of the heaviest horses imperative. The streets had become congested. It took as long to move a ton as it did 5 tons—or 7 tons, for that matter—simply for the reason that progress beyond a certain fixed slow rate of speed had become impossible.

"The International offered prizes for horses of stated weights all of them beyond the generally accepted standard—the campaign for better big geldings was set in motion.

"The Packers and other users of large numbers of horses decided that it was up to them to use more power.

"A furor at once developed. Brewers, teaming contractors, paper makers and dozens of other kinds of merchants grasped the idea in a minute. The heavy horses were to be found. The streets were congested. The heavy horse was the thing. And right in these

Care of the Draft Horse.

Following are the drivers' rules of the Boston Work Horse Parade Association:

1. Start at a walk and let your horse work very easily for the first half hour.

2. A heavy draft horse should never be driven faster than a walk, with or without a load.

3. Look to your harness. Avoid these faults especially:
Bridle too long or too short.
Blinders pressing on the eyes or flapping.
Throat latch too tight.
Collar too tight or too loose.
Traces too long.
Breeching too low down or too loose.

4. Drive your horse all the time. Feel his mouth gently. Never jerk the reins.

5. Take the horse out of the shafts as much as possible; and if you drive a pair of four, unfasten the outside traces while the horses are standing; they will rest better that way.

6. Teach your horses to go into the collar gradually. When a load is to be started, speak to the horses and take a firm hold on the reins so that they will arch their necks, keep their legs under them and step on their toes.

7. Water your horse as often as possible. Water in moderate quantities will not hurt him, so long as he keeps moving.

8. Blanket your horse carefully when he stands, especially if he is at all hot. Repeated slight chills stiffen and age a horse before his time.

9. Bring your horse in cool and breathing easily. If he comes in hot he will sweat in the stable, and the sudden stopping of hard work is bad for his feet.

10. In hot weather or in drawing heavy loads, watch your horse's breathing. If he breathes hard, or short and quick, it is time to stop.

11. Remember that the horse is the most nervous of all animals, and that little things annoy and irritate him. Remember that he will be contented or miserable accordingly as you treat him.

Credit: Author's Collection

"CARE OF DRAFT HORSES" was published in the October 1913 issue of "The Blacksmith and Wheelwright" magazine.

THE HORSES

few words is contained the history of the rise of the drafter to an appreciation of his true merit in the country.''

''The City of Chicago passed an ordinance—'It is unlawful for a vehicle drawn by one horse or mule to carry a load exceeding 3,500 pounds or for one drawn by more than one horse or mule to be loaded above 4,000 pounds for each animal hitched to the vehicle.

''Besides this, any wagon will be considered overloaded when it is evident its weight is beyond the capacity of the horses to handle it without overtaxing their strength.

''The effect of this ordinance added to the traffic conditions.

''Teaming concerns are now using every horse it can press into service. Current rate for hired teams is $8.00 a day for a pair and a driver and $3.00 for each additional horse.'' (*Breeder's Gazette* Jan. 5, 1910)

''We have 150 horses now in the stables of Armour & Co. as our use of the motor truck

PAUSE THAT REFRESHES. After a hard day's work in the fields, these Percherons drink deep.

REST AREA. One of the watering stops provided for draft animals in the city offers an oasis for a thirsty laborer.

COOL WATER. Refreshing themselves on a summer day in Michigan, draft horses enjoy the water pumped up by the windmill in the background.

FRIENDS. Many milk wagon horses working the same route day in and day out became pets of neighborhood kids. Here the boys offer a couple of lumps of sugar to an interested horse on a Milwaukee route.

CURES AND REMEDIES. These ads from a 1908 *Breeder's Gazette* offer products for strained tendons, wind puffs, ringbone, distemper, spavin, heaves, indigestion, gall, swollen glands and other equine problems.

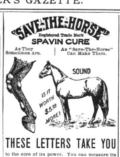

BACK UP. Part of the education of any draft horse is to learn how to back up a load. This is particularly true with teams used in hauling freight in cities. This team is backing a platform dray wagon up to the Pere Marquette freight docks in Milwaukee.

THE HORSES

BACK SCRATCH. After working all day, particularly a hot one, horses loved to roll if given an opportunity.

has greatly reduced our need for horse labor," wrote Armour's Thomas Donnellan in the October 29, 1914, *Breeder's Gazette.* He continued, "Our horses weigh from 1,300 pounds to 2,000 pounds and most of them are high grades. We have to be constantly renewing our supply as the number is being constantly reduced by losses from various causes. We figure on a loss of 12 a year due to horses wearing out, meeting with accidents or dying from a fatal malady. We also sell a horse now and then when we think he will be of more value on a farm than with us. Most of the horses we buy are geldings.

"Our feed consists chiefly of timothy hay and crushed oats. We give a soft feed twice a week consisting of chopped hay, bran, and ground oats. This mixture we cook by steam in a big vat. We do not feed much corn for if fed too heavily to working horses it produces colic. It is too heating. We cannot feed these hard working horses, as we would a stallion. They have to make 25 miles a day and cannot do that on soft feed. We have also found that when we feed soft feed in considerable quantities and quite continuously the horses get colic. When, however, we want to fit horses for the show we feed them largely on ground feed and clover hay.

"Good timothy hay and good oats are two important feeds for keeping work horses in good condition."

Encouraging Better Treatment

Work horse parades in cities became a subtle and quiet way to try and educate

Hot Weather Rules for the Horse.

Load lightly, and drive slowly. Stop in the shade if possible. Water your horse as often as possible. So long as a horse is working, water in moderate quantities will not hurt him. But let him drink only a few swallows if he is going to stand still. Do not fail to water him at night after he has eaten his hay. When he comes in after work, sponge off the harness marks and sweat, his eyes, his nose and mouth, and the dock. Wash his feet but not his legs. If the thermometer is 75 degrees or higher, wipe him all over with a wet sponge. Use vinegar water if possible. Do not turn the hose on him. Saturday night, give a bran mash, cold; and add a tablespoonful of saltpetre. Do not use a horse-hat, unless it is a canopy-top hat. The ordinary bell-shaped hat does more harm than good. A sponge on top of the head, or even a cloth, is good if kept wet. If dry it is worse than nothing. If the horse is overcome by heat, get him into the shade, remove harness and bridle, wash out his mouth, sponge him all over, shower his legs and give him four ounces of aromatic spirits of ammonia, or two ounces of sweet spirits of nitre, in a pint of water, or give him a pint of coffee warm. Cool his head at once, using cold water, or, if necessary, chopped ice, wrapped in a cloth. If the horse is off his feed, try him with two quarts of oats mixed with bran, and a little water; and add a little salt or sugar. Or give him oatmeal gruel or barley water to drink. Watch your horse. If he stops sweating suddenly, or if he breathes short and quick, or if his ears droop, or if he stands with his legs braced sideways, he is in danger of a heat or sunstroke and needs attention at once. If it is so hot that the horse sweats in the stable at night, tie him outside. Unless he cools off during the night, he cannot well stand the next day's heat.

THIS PIECE, "Hot Weather Rules For The Horse," appeared in the August 1914 issue of "The Blacksmith and Wheelwright" magazine.

LUNCH BREAK. The tie weight, sometimes called a "hitching" weight, is covered with slush. This 20-lb. cast-iron weight, snapped to the bridle, kept the horse from wandering off. Note that the bit has been removed from the horse's mouth while the nosebag is on.

COLD WEATHER WORKERS. The temperature was zero as the hot breath of the horses eating their midday meal steamed up out of the dinner pails. On cold days like this horses were usually blanketed.

LUNCHTIME. The near horse is looking back anxiously expecting his noontime bait of oats the driver is pouring into a pail. Teamsters who knew they would be away from the stable all day carried grain with them.

teamsters to give their horses better and more considerate care.

"Work Horse Parades in cities grow—San Francisco recently had one of draft, express and delivery horses and mules.

"There were 2,069 horses besides 300 Police horses directing the march. All horses were groomed with painstaking care, polished harness and shining brass.

"Discerning horsemen saw the scarcity of really choice horses. Also noted was a lack of teamsters who have the ability to preserve the vigor and bloom of their horses doing hard work. Consequently, between poor natural endowment and poor care, most horses in city service have many prominent faults." (*Breeder's Gazette* Sept. 29, 1909)

This magazine again reported in its January 19, 1910 issue, "Boston Work Horse Parade Association in addition to sponsoring the work horse parade, has developed a stable competition to every kind—livery, hack, boarding. Prizes will be awarded.

"The stables entered are judged not in competition with one another, but according as to how they satisfy the judges.

THE HORSES

"Among the points to be considered by the Judges will be the quality of the hay and grain, bedding, watering, blanketing, grooming, ventilation, stalls, sanitary conditions of the stable and handling of the horses by the drivers and grooms. The costliness of the stable and its furnishings will not count. The intention is to give the poor owner an equal chance with the rich owner.

"A detailed report of the condition of each stable will be made only to its owner and the Judges are instructed to offer him such suggestions as may remedy the defects found."

The City of Chicago took an interest in the value of such parades. The *Breeder's Gazette* for April 6, 1910 commented, "Chicago Work HORSE Parade Organized. Horses will be classified according to work they perform. Their type, condition, age, suitability and fitting of harness and manners as workers will all be considered. The aim will be to encourage better care, selection and management.

"The foundation for successful competition must be good horses."

Two weeks later the Gazette reported, "Chicago parade set for May 30th, Decoration Day—everything from a billposters rig to a packer's six-horse team will be shown.

"Over 1,000 entries to date have been received.

"The event will take place on Michigan Ave.

"A new harness, or new wagon, counts for nothing.

"The horses only are to be judged.

"Any horse that is dock-tailed, sick, lame, thin, galled, out of condition or otherwise un-

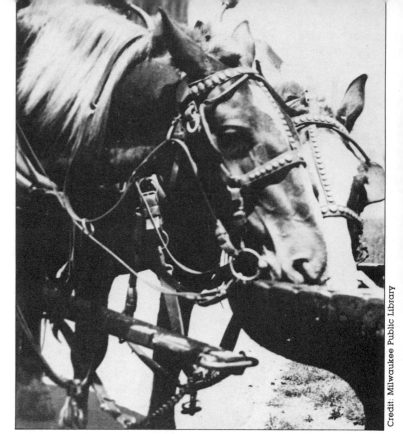

TIME FOR A DRINK. Cities maintained public watering fountains scattered all over town at main intersections where the working horses could get cool, fresh water.

ICICLES. On sub-zero days the hot breath of the horses would form icicles on their noses. Thoughtful drivers put their warm hands over the horse's noses periodically to melt and remove the ice deposits.

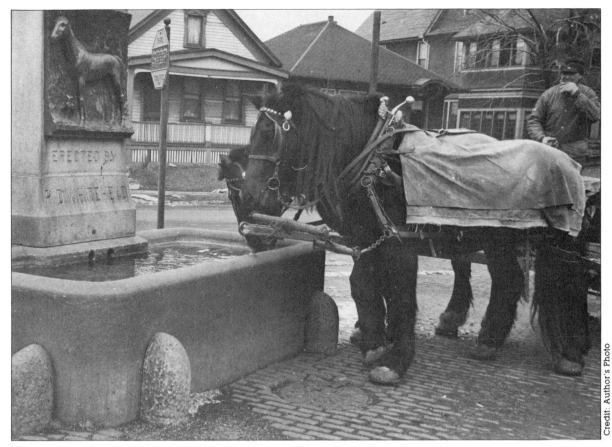

ON MILWAUKEE'S SOUTHSIDE. Occasionally, a public-minded citizen, who had a love and concern for horses, would contribute to the erection of a water fountain.

OASIS FOR HORSES. Henry Bergh was the founder of the American Society for the Prevention of Cruelty to Animals. In 1891 the city of Milwaukee erected a handsome monument to Mr. Bergh directly in front of City Hall. This view, looking south, shows the statue surrounded by a watering fountain for horses. Two horses are just leaving, one just arriving, and one drinking.

THE HORSES

HORSE AMBULANCE. The Boston Fire Department maintained this vehicle to take care of emergencies that might involve injured animals.

fit for work, will be excluded.

"There will be special classes for old horses, contractors, runabout horses, letter carriers' horses, horses owned by cities, towns, and fire departments.

"Docile and gentle manners will be considered showing that the horses have been unkindly treated.

"Harness must be clean and comfortable, well-fitting and not unnecessarily heavy and collar not too small or large. Harness that is light, but strong enough to do the work required will be preferred to heavier harness.

"Wagons will not be considered except that a vehicle too heavy for the work in which horses are used will be disqualified.

"The Work Horse Parade Association desires to take special notice of old horses in Cook County and deems this class most important in the parade. Class is open to horses who have seen at least 10 years service—years of service is not only important, but horse must look well fed, sleek and comfortable and in good condition.

"The Parade is expected to improve horses and their treatment. Many drivers treat their horses as though they were a machine. On the other hand, there are many humane teamsters who have a real affection for their horses and take the greatest pride in their appearance.

"To reward and increase this class is an object of the Chicago parade".

A week before the big Chicago event, the *Breeder's Gazette* had this to say, "Chicago Work Horse Parade. Every harness and paint shop in the city has been swamped with rush orders for the past month.

"Although elegance of equipment will count for nothing in deciding awards, there is no disposition to send the work horses out on a gala occasion without shining harness and glistening wagons.

"Probably $60 million is represented by Chicago's 250,000 horses, and as much more capital is invested in barns, harness and wagons.

"Many managers of horses confess they did not realize that there was any economical way to lengthen the life and increase the working ability of their horses until instances of long service began to develop among the entries of this parade."

In the June 1, 1910 issue, the *Breeder's Gazette* said this about the big parade, in addition to listing all the winners in all classes:

"Chicago Work Horse Parade May 30—Everyone was astonished to see such a uniformly superior lot of work horses. There were 1,500 entries, 1,000 wagons and teams from one horse to eight horses, 2,500 horses altogether. It took one hour and 40 minutes for the parade to pass, and at that, they were in double file.

"If strung out single file they would stretch out eight miles and take three hours to pass."

The value of these work horse parades in the big cities continued to increase. The *Breeder's Gazette* for July 2, 1913, said, "In cities where work horse parades have been held regularly for a series of years, a perceptible improvement may be noted in the character and condition of the teams exhibited. When city work horses are gathered together there is a keen rivalry. In the recent Boston parade there were 1,231 entries and 1,516 horses."

A year later Boston had 1,700 entries in their work horse parade, and the *Breeder's*

TIME FOR A TREAT. This horse's attention is not on his job of pulling a bakery wagon in Troy, New York. He wants his usual tidbit of a piece of apple, or carrot, or dry bread.

THE HORSES

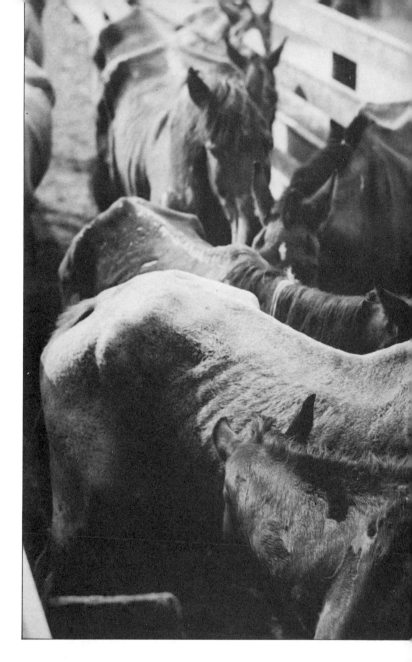

Gazette of June 11, 1914, reported that ''this city has the Boston Work Horse Relief Association to encourage owners and drivers to treat their horses well in the stable and on the street.

''A free hospital has been established for horses belonging to poor men, but for any horse in a case of an emergency.

''The Association carries on stable inspections and defects are remedied.

''Smoke fests for drivers are held from time to time at which experts give advice on care and management.

''The Association distributes hot weather rules, stable rules, driver's rules, as well as posting them in stables.''

TO A HORSE

O HORSE, YOU ARE A WONDROUS THING!
NO HORNS TO HONK, NO BELLS TO RING;
NO LICENSE BUYING EVERY YEAR
WITH PLATES TO SCREW ON FRONT AND REAR.

NO SPARKS TO MISS, NO GEARS TO STRIP;
YOU START YOURSELF, NO CLUTCH TO SLIP,
NO GAS BILLS MOUNTING EVERY DAY
TO STEAL THE JOY OF LIFE AWAY!
YOUR INNER TUBES ARE ALL O.K.
AND THANK THE LORD, THEY STAY THAT WAY.

YOUR SPARK PLUGS NEVER MISS OR FUSS;
YOUR MOTOR NEVER MAKES US CUSS.
YOUR FRAME IS GOOD FOR MANY A MILE
YOUR BODY NEVER CHANGES STYLE,
YOUR WANTS ARE FEW AND EASILY MET
YOU'VE SOMETHING ON THE AUTO YET.

WORN OUT. The peddler who owned this pitifully decrepit horse had no feeling for the animal. Thin, gaunt and crippled, the horse hardly existed on its meagre rations. When touched by the ever present whip, the horse would plod along.

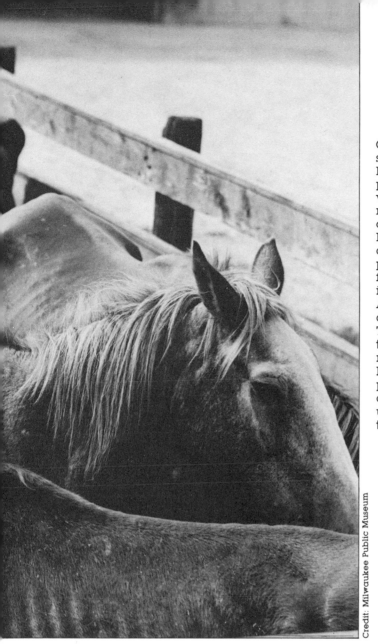

OLD AGE. The only crime committed by the horses shown here was that of growing old and, perhaps, going lame because of constant pounding on brick pavements. Is one of these horses remembering years past when he galloped down the street pulling a shiny red fire engine? Is another remembering prancing through a tree covered residential neighborhood with his polished harness and handsome wagon delivering orders? Is another remembering trotting down a country road pulling a surrey taking the master and his family to a church picnic? Or, are these horses so broken in spirit that they don't remember and don't care? As a one-time powerful and hardworking horse began to decline because of old age, it changed owners. The more "stove up" the horse became, the more often it changed hands. Even the horse's name was lost in the shuffle. Too few owners had the compassion, or the facilities, to put old horses out to pasture. As the loyal animals wore out they could do less work and this only encouraged their downhill slide until at last they were so ganted-up and hidebound that they were only good for dog feed.

A SAD END. The wornout horse has terror in its eye from the strange surroundings and strange smells of the abattoir. All the years of faithful service is wiped out coldly and callously with the crack of a high-powered pistol.

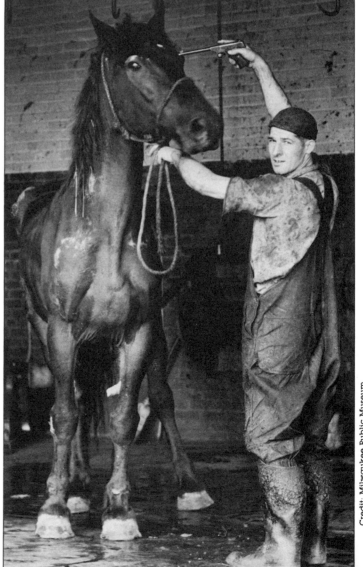

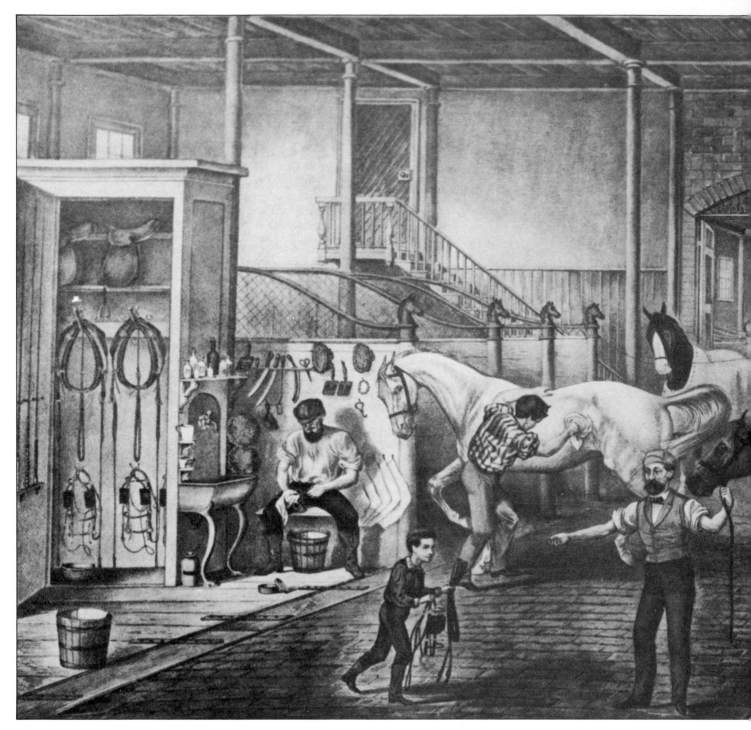

FULLY EQUIPPED STABLE. This Currier & Ives print could illustrate a private stable or a livery stable. Cleaning harness, rubbing down horses and hitching horses creates a lot of activity. On the right is a group of buggies and above them, in storage for the summer, are the cutters.

FAN OF HORSEPOWER. H. J. Heinz, founder of the company that bears his name, is seen here with one of his sons in front of the stable at Greenlawn, where his mansion was located in Pittsburgh's east end. Mr. Heinz said, "It was a sad day for mankind when the internal combustion engine replaced the horse."

Credit: Library of Congress

Credit: Author's Collection

CHAPTER X

THE STABLES

The CARDINAL RULES for a good horse stable: it should be dry, have good ventilation and be free of drafts.

Packed clay was the best flooring for a horse, but most stables were floored with planks, which were also excellent. Concrete or asphalt were never suggested.

For draft horses a tie stall was a minimum of four feet wide, preferably five feet wide, no more. Lighter weight driving horses required less space.

An array of equipment was required in a horse stable, including brooms, pails, shovels, hay forks, manure forks, brushes, squeegees, curry combs, mane comb, clippers, and hoof pick. Other stable accessories were blankets, sponges, axle grease, wheel jack, whips, soap and harness.

Private stables were always neat, clean and convenient. Frequently, the larger the stable the fancier it was. The coachman in a private stable managed the operation and he usually had one man for every four horses. This crew had to clean stalls, shake out blankets, clean harness, vehicles, bits and chains, and groom, feed and water the horses.

Many affluent companies took particular pride in their stables and their horses. The H.J. Heinz Co. stabled 110 jet black horses. At work they were hitched to white wagons

---

DELUXE STABLE. Varnished paneled ceilings, screened windows, fancy lamps and a cork floor added to the good looks of this stable of a wealthy gentleman in the 1890s. Note the stone watering fountain in the aisle.

175

THE STABLES

trimmed in green. The stable was a fireproof structure, steam-heated, lighted by electricity, with screens on all the windows. The floor was cork brick covered with sawdust.

In the wintertime an attentive teamster always checked the horses' shoes before going out on the ice and snow to be sure all caulks were in place. A thoughtful teamster always warmed the cold bit in his hand before inserting it in the horse's mouth.

DELUXE ACCOMMODATIONS. H.J. Heinz Co. of Pittsburgh maintained a fireproof, steam-heated stable for 110 draft horses. Note the trolley to carry the harness to and from the horses. Most of the big companies maintained excellent stables and gave their horses fine care and attention.

ORDERLY STABLE. Another floor of the Heinz stable. The horses were evenly matched in size and always well-groomed.

WATER THERAPY. In the Heinz stable one box stall was fitted out with short concrete walls. This then became a water-pit where a horse that developed sore feet from working on paved streets could stand in water up to its pasterns to draw out the soreness and soften the feet.

TIME TO REST. After a day of hauling lumber around the yard, this horse heads into the stable. Note the long winter coat on this Belgian gelding.

OLD-TIME DRIVER. Lefty Walsh drove horses from the age of 13, and said he hoped he never had to lower himself to drive a truck. Here he has unhitched his team from a freight wagon and is leading the "off horse" into the barn. The "near horse" follows.

SCHLITZ STABLES. The facade of the stable building of the Jos. Schlitz Brewing Co. on Walnut Street in Milwaukee. Note the handsome stone horse heads set into the brick wall.

ROUTE WAGON. Work is done for the day and the team heads into the stable of Milwaukee's Gridley Dairy Co. The wagon will be washed and the horses curried, brushed, fed, and watered.

AT DAY'S END. The Fifth Street stables of the Luick Dairy Co. in Milwaukee. The well-brushed and curried horses have finished their routes for the day.

HANGING HARNESS. Each horse's harness hangs on a large hook attached to the posts dividing the stalls. Many operations preferred keeping harness in separate rooms to keep it from being exposed to the urine and manure fumes.

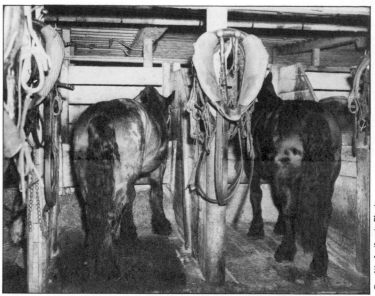

EASY ON THE FEET. Plank flooring as used here, seemed best for horses' feet. Concrete was hard on the animals.

ANOTHER VIEW. The pair of blue roans in the foreground had a wholesale route and pulled a big wagon.

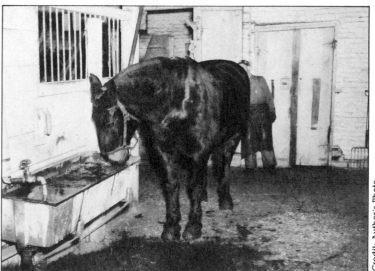

PAUSE THAT REFRESHES. Every good stable had a water tank, so a horse could get a drink after eating his allotment of hay.

JANESVILLE, WIS., C-1900. Those who could afford it had a small barn in back of their homes as seen here. Vehicles were stored, and one, two or perhaps three, horses were stabled.

TALLMAN HOUSE, JANESVILLE, WIS. This mansion had a beautiful stable out back that matched the house in style and elegance. Most every barn had a hay loft. The large second floor door in the stable is for hay.

WAUKESHA, WISCONSIN. The stable facilities for this handsome home can be seen in the background. The barn would provide storage for carriages and sleighs, as well as stalls for the horses used by the family. Note the hitching post at the curb.

FINELY DECORATED. The elegance and trimmings of this stable indicate it served an extremely wealthy family.

EQUINE COMFORT. Well-swept floor, deep bedding of straw, and the constant attention of a stable boy were the earmarks of a topnotch stable.

ORNATE HOUSEHOLD. The stable, seen in the rear of the mansion, was an important part of the operation. The building contained a hay mow, stalls for horses, carriage storage room, harness room, wash room, feed room and possibly living quarters for help. This home (C-1890) was owned by Frank K. Bull in Racine, Wisconsin.

BUILT IN 1903. The stable behind the mansion of J. I. Case, farm machinery manufacturer of Racine, Wisconsin.

INTERIOR VIEW. The J. I. Case stable was elegantly appointed with white tile walls and mahogany woodwork. The horses, bedded deep in straw, are all blanketed.

ONE FAMILY'S VEHICLES. A carriage for every occasion. This selection shows the neat, clean way these vehicles were cared for.

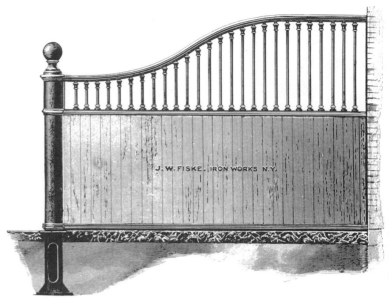

STALL PARTITIONS. The catalog of J. W. Fiske Iron Works offered these.

SPACIOUS COACH HOUSE. Stables at the Gedney Farms in White Plains, New York, around 1900 had 20 box stalls, 10 standing stalls, a hitching room, coach house (shown here holding 20 carriages), the owner's room, harness room and cleaning room. The servants' rooms and manager's apartment were on the second floor of the coach house.

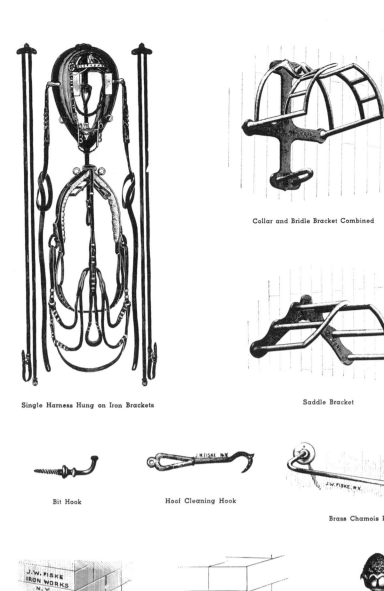

Single Harness Hung on Iron Brackets

Collar and Bridle Bracket Combined

Collar Bracket

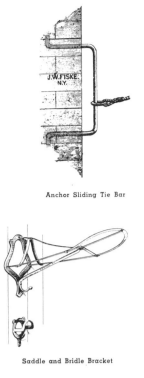

Anchor Sliding Tie Bar

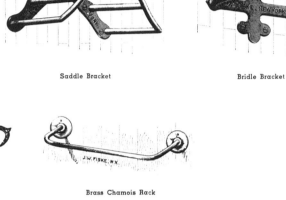

Saddle Bracket

Bridle Bracket

Saddle and Bridle Bracket

Bit Hook

Hoof Cleaning Hook

Brass Chamois Rack

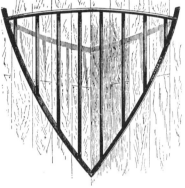

Cast Iron Wheel Guard

Expansion Bolt Tie Ring

Pineapple Head for Post

Anchor Tie Ring

Wrought Iron Hay Rack

Horse Head for Wood Post

Post Cap for Wood Post

Wrought Iron Box Stall Hinge

Plain Cap for Wood Post

Post Cap

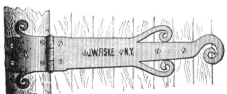

Harness Bracket

STABLE HARDWARE. Many varieties were illustrated and offered by J. W. Fiske Iron Works.

HITCHING POST. Cast iron horse heads, fastened to wood posts, were popular for tying horses. They were a necessity in front of residences and business establishments alike. In western towns, tie racks that could accommodate six or eight horses—or more—were commonplace.

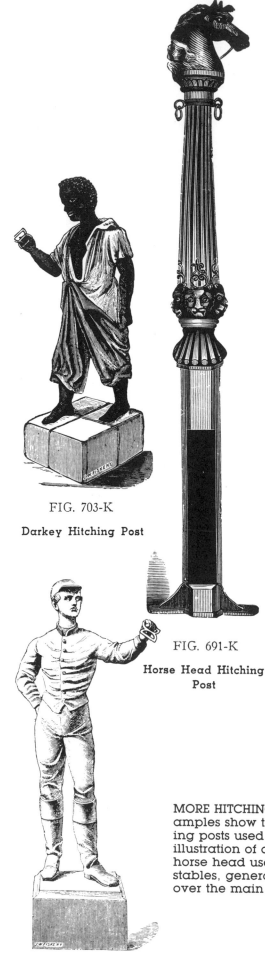

FIG. 703-K

Darkey Hitching Post

FIG. 691-K

Horse Head Hitching Post

FIG. 693-K

Pineapple Hitching Post

FIG. 705-K

Jockey Hitching Post

MORE HITCHING POSTS. A few examples show the variety of hitching posts used. At right an illustration of a cast zinc life-sized horse head used to decorate stables, generally on the outside over the main door.

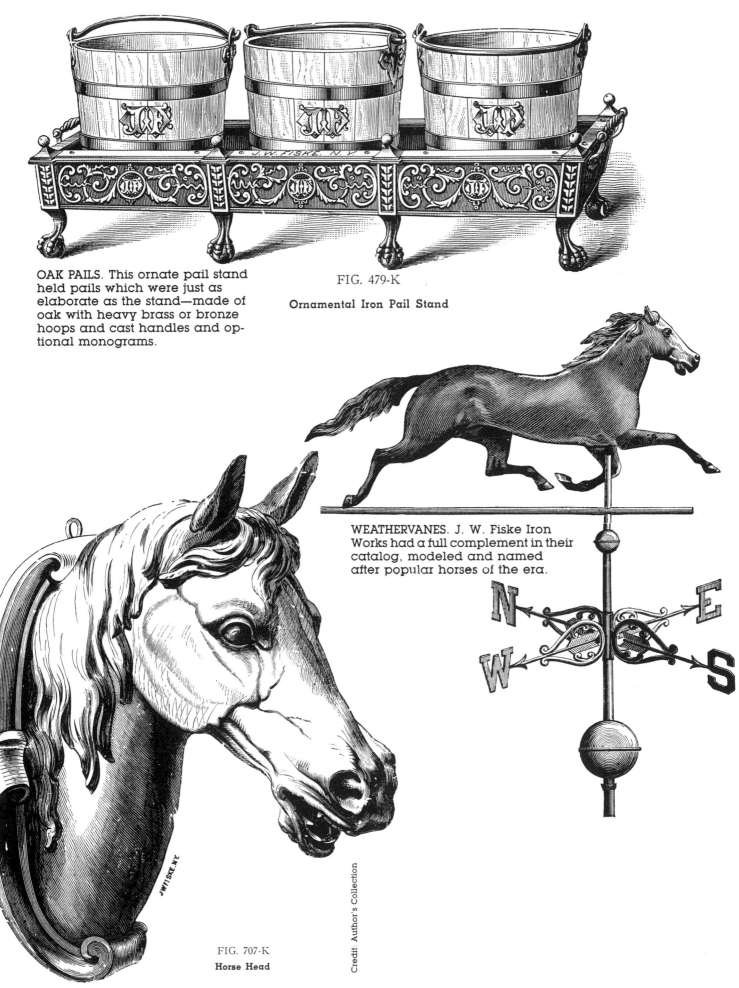

OAK PAILS. This ornate pail stand held pails which were just as elaborate as the stand—made of oak with heavy brass or bronze hoops and cast handles and optional monograms.

FIG. 479-K

Ornamental Iron Pail Stand

WEATHERVANES. J. W. Fiske Iron Works had a full complement in their catalog, modeled and named after popular horses of the era.

FIG. 707-K

Horse Head

Credit: Author's Collection

CHAPTER XI

THE HARNESS AND TRAPPINGS

IN THE HEYDAY of horse power, there were scores of harness makers, from huge factories, producing carloads, to individuals in small towns, cutting and sewing one harness at a time.

There were many designs and variations —harness for every conceivable type of work and load and every size of horse.

About their harness, Sears, Roebuck & Co. claimed, ''You can't find its equal for less than $60.00.'' Walsh Harness Co. of Milwaukee, featured harness designed without the use of buckles, giving it a longer life.

For the carriage trade, harness was available made from black patent leather with all hardware either nickel- or brass-plated.

Horse owners all realized that proper care

FOR ALL YOUR HARNESS NEEDS. The sign is not yet hung over the door, but the harness shop is open for business. Harness is hung on a hook and placed in the window. Lap robes and horse blankets are displayed, as well as a box exhibiting a large selection of buggy whips.

HARNESS SHOP

THE HARNESS AND TRAPPINGS

and treatment of a leather harness would maintain its suppleness and greatly extend its life. The harness had to be cleaned and sponged of sweat and grime with a good saddle soap. It was hard work, but the results showed on the rig as it moved down the street.

"Cleaning Harness In Spring" was the title of an article in the March 12, 1914, issue of the Breeder's Gazette. It offered this advice: "It is remarkable how much soreness of neck, shoulder, breast and thigh can be avoided by properly softening and cleaning those portions where the strains of severe draft are felt.

"The preparation of the harness involves the cleaning, the oiling, blackening and the adjusting.

"Cleaning the leather of the accumulated sweat, skin scurf, dust and hairs removes many roughened sources of irritation, makes the harness cooler and permits better absorption of the oil and stain.

"It is common practice to scrape off the accumulations of work with the edge of a dull knife, but a more satisfactory method, conducive to longer life of the harness, is to soak up the dirt with a watery lather of soft soap. After standing long enough to wet the dried accumulation a brush with medium stiff bristles will remove the dirt without injuring the surface of the leather. The soap should then thoroughly be rinsed off, the harness allowed to become nearly dry and the blackening solution applied. When this is dry the oiling may take place and a double application of oil is always recommended since the second dose adds much to the pliability."

The fortunes of harness makers skidded as the horse began to lose out to cars and trucks. Small companies dropped out of the harness

REALISTIC DISPLAY. Thomas C. Smith Harness Shop in Milwaukee (1888) had its dressed-up dummy horse right out front.

Credit: Milwaukee Journal

LIFE-SIZED. Dummy horses made from papier-mache were used to show off saddles and harness in life-like fashion—sort of like mannequins in dress shop windows. The Standard Harness Shop on Grand Ave. (C-1899) Milwaukee, Wisconsin, was no exception.

BRASS HARNESS. $30.49 AND $37.29

FINE BRASS TRIMMED FARM HARNESS

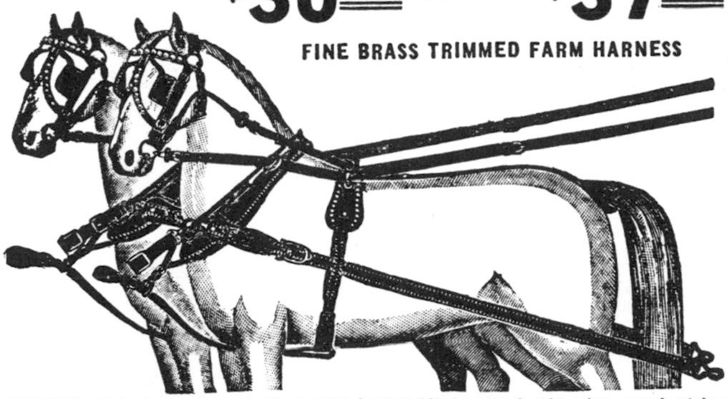

BRIDLES—¾ inch, box loop cheeks, patent leather blinds, round side reins, round winker stays, brass spotted face piece.
LINES—1 inch, 20 feet long, with buckle and billet with snap.
HAMES—Low top wood hames, brass long spot, Hayden holdback ring, four hame straps and two spread straps, box loop, scalloped, hame tug and Champion trace buckle.
TRACES—1½ inches wide by 6 feet 6 inches long, raised center, round edge finish, cockeye sewed in, to buckle in hame tug.
PADS—Harness leather sewed bottom, and stuffed with hai., brass spots all around pad. Flexible tree pad, single strap harness leather skirts, double and stitched raised bearer, bellyband folded, 1-inch back straps, scalloped and stitched with crupper to buckle on.
BREAST STRAPS—1½ inches, with snaps and slides, adjustable.
MARTINGALES—1½ inches, with collar straps, adjustable.
TRIMMINGS—Brass swage trimmings.
MATERIAL used in this harness is the very best grade, the best tanned, the best selected Dundee oak leather. All the straps of this harness are carefully cut and selected for weight of strap. The straps are carefully edged and blacked and the harness is smooth finished, using the best brass hardware. The workmanship is first class and together with the excellent quality of leather and trimming, makes it one of the best harness we have to offer for $30.49. Harness similar to this in description, but not as good in quality, are sold for $8.00 to $12.00 more than our price.
Weight of harness, packed for shipment, about 80 pounds.
No. 10K825 Price, brass trimmed, 1½-inch traces, without collars **$30.49**
No. 10K826 Price, brass trimmed, 1½-inch traces, with double hip strap breeching, without collars **$37.29**
Add extra for full brass iron hames in place of wood hames **2.50**
Add extra for good ¾-inch tie straps, each .... **.26**
Traces without cockeye will be furnished at same price if wanted.
For prices on collars see collar page.

WILLIAMS' SPECIAL HOOK AND TERRET FARM HARNESS

No. 10K820	No. 10K823
$17.59	$18.59
1½-INCH TRACES.	1¾-INCH TRACES.

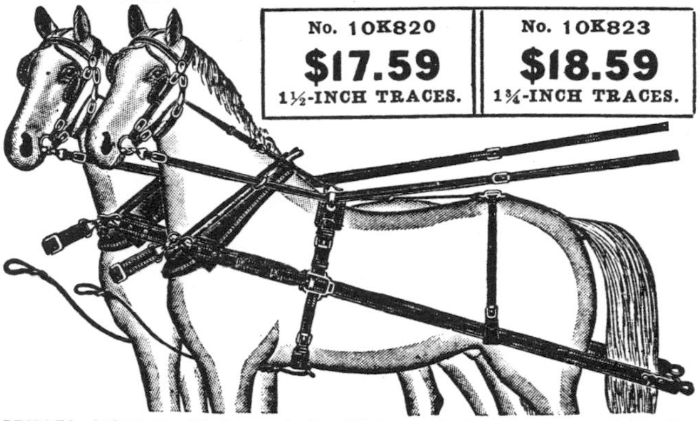

BRIDLES—¾ inch, sensible harness leather blinds, round winker braces, round side reins, heavy crown piece.
LINES—⅞ inch by 18 feet long, with snap.
HAMES—Wood hames, iron bound over top, XC plated with Hayden holdback ring; hame tugs, folded harness leather, two leather loops, 1½-inch layer double and stitched with Champion trace buckle sewed in. Four hame straps, two spread straps.
BREAST STRAPS—1½ inches, double with snaps and slides.
TRACES—1½ inches and 1¾ inches wide by 6 feet long, double and stitched with clip cockeyes riveted on.
PADS—All harness leather flexible pads, with hooks and terrets, 1¼-inch double market strap skirts, to buckle in the hame tugs; 1-inch back straps and hip straps sewed in. Cooper trace carrier, folded crupper to buckle on, bellybands folded.
MARTINGALES—1½ inches.
TRIMMINGS—Full XC white metal.
Weight of harness, packed for shipment, about 65 pounds.
No. 10K820 Price, with 1½-inch traces, without collars **$17.59**
No. 10K823 Price, with 1¾-inch traces, without collars **18.59**
Add extra for team breeching **$3.25**
Add extra for 1 inch by 18 feet lines. **.50**
Add extra for extra large harness for 1,400 to 1,700-pound horses **3.25**
Add extra for collar strap ..... **.30**
Add extra for ¾-inch tie strap, each .... **.26**
For prices on collars see collar page

REDUCED IN PRICE FROM $18.38 TO $17.39 AND REDUCED IN PRICE FROM $19.38 TO $18.29

FARM HARNESS.

BRIDLES—⅞-inch cheeks, sensible blinds, flat winker stays, short reins to run over the hames.
LINES—⅞ inch wide by 15 feet long, with snaps.
HAMES—Common wooden hames, oval iron pattern with holdback ring, hame tugs folded with 1½-inch layer and Champion trace buckle, hame strap and spread strap.
TRACES—1½ inches and 1¾ inches, double and stitched, 6 feet long with clip cockeyes riveted on to buckle in the hame tug.
PADS—Folded body, flat layer, 1¼-inch billets, folded and stitched bellyband, 1¼-inch back straps, 1-inch hip straps, 1½-inch breast straps, with snaps and slides riveted.
MARTINGALES—1½ inches, no collar straps.
TRIMMINGS—XC white metal.
Weight of harness, packed for shipment, about 58 pounds.
No. 10K804 Price, with 1½-inch traces, without collars **$17.39**
No. 10K808 Price, with 1¾-inch traces, without collars **$18.29**
Add extra for team breeching **3.20**
Add extra for ¾-inch tie strap, each, **.26**
For prices on collars, see collar page.

THE LOWER PRICES
QUOTED IN THIS CATALOGUE will be more interesting to you than the prices quoted by any other firm or individual.

$38.97 EASTERN HARNESS.

BRIDLES—⅞ inch, box loop cheeks, square patent leather blinds, spotted face pieces, fancy front, round winker stays, brass rosettes, solid crown piece, heavy throat latch and flat rein.
LINES—1 inch wide, 15 feet long, buckle and billet with snaps, extra heavy and well made.
HAMES—Brass long spot, ball top, iron clad, painted red, clip and staple, four hame straps and two spread straps.
BREAST CHAINS—Heavy twisted link chain with T bar and snap on one end.
MARTINGALES—1½-inch from collar to bellyband, heavy buckle and billet to buckle around the collar and loop for bellyband.
BREAST STRAPS—1½ inches, double and stitched, with snaps and slides.
TRACES—2 inches wide, 4 feet 6 inches long with three rows stitching and dee loop in center, making a sectional trace, 3½-foot trace chain to snap in the ring.
PADS—Swell housing, 1½-inch layer double and stitched the full length, brass spots, 1½-inch billet to buckle in the lead up from dee in trace, heavy folded bellybands.
BREECHING—Heavy folded harness leather body, 1½-inch layer, double and stitched the full length, heavy rings, 1¼-inch double hip straps, 1½-inch back straps running to the hame, 1¼-inch side straps from dee in trace to ring in the breeching, large padded safe under the ring on the hips.
TRIMMINGS—Brass.
Weight of harness, boxed for shipment, about 90 pounds. We do not make any changes in this harness.
Furnished in complete set for two horses.
No. 10K816 Price, of harness without collars **$38.97**

$21.99 SHORT TUG FARM HARNESS.

BRIDLES—⅞ inch, double and stitched short cheeks with noseband, heavy double harness leather blinds stitched together, flat winker braces, leather front, solid crown piece and heavy throat latch, flat rein.
LINES—1 inch wide, 18 feet long, with buckle and billets with snap at the bit.
HAMES—Red ball top clip and staple, four heavy hame straps and two spread straps.
BREAST CHAINS—Heavy twisted link chain with T bar and snap.
MARTINGALES—1½ inches, extra heavy to buckle around the collar and loop for bellyband.
TRACES—1¾ inches, 4 feet 3 inches long, double and stitched, 3½-foot chain to snap in the ring; an extra heavy well made trace.
PADS—Flat folded body with 1½-inch layer, 1½-inch billets to buckle around the trace, 1¼-inch bellyband billets, heavy folded bellyband.
BACK STRAPS—1½ inches running to the hame, 1-inch hip straps to buckle in the ring on the trace, folded crupper to buckle on.
TRIMMINGS—All trimming black japan finish with ball top hames.
Weight of harness boxed for shipment, about 50 pounds. We do not make any changes in this harness.
Furnished in sets for two horses.
No. 10K814 Price of harness without collars **$21.99**
Add extra for 1¾-inch traces **1.00**

HERCULES FARM HARNESS

REDUCED FROM PREVIOUS CATALOGUE. **$20.99 AND $21.99**

SEE THAT PAD, the best Flexible Tree, on sewed and stuffed leather bottom. This makes the best flexible pad.

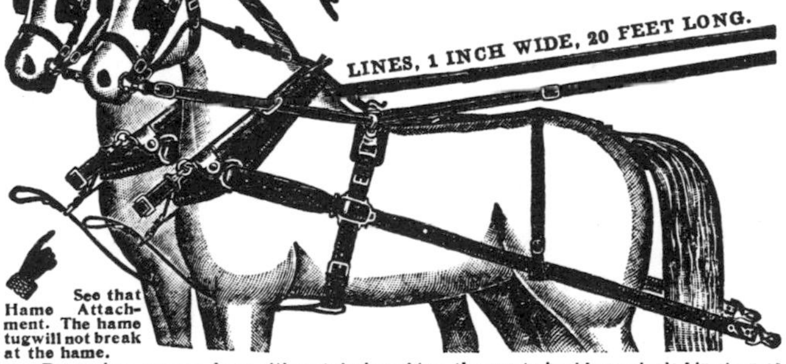

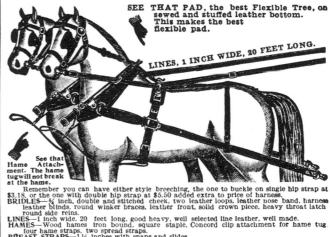

See that Hame Attachment. The hame tug will not break at the hame.

Remember you can have either style breeching, the one to buckle on single hip strap at $3.18, or the one with double hip strap at $5.50 added extra to price of harness.
BRIDLES—¾ inch, double and stitched cheek, two leather loops, leather nose band, harness leather blinds, round winker braces, leather front, solid crown piece, heavy throat latch round side reins.
LINES—1 inch wide, 20 feet long, good heavy, well selected line leather, well made.
HAMES—Wood hames iron bound, square staple, Concord clip attachment for hame tug, four hame straps, two spread straps.
BREAST STRAPS—1½ inches with snaps and slides.
MARTINGALES—1½ inches with collar strap, ring in the loop.
HAME TUGS—Folded body, 1½-inch double and stitched layer, double loop attached to the hame with Concord clip hame attachment, making a jointed bolt hame tug, will not break at the clip, Champion trace buckle, 1¼-inch bellyband billet, heavy folded bellyband.
PADS—Improved Moline tree, flexible joint, harness leather bottom, double and stitched, hair stuffed, single strap skirts with 1¼-inch layer doubled and stitched, to buckle in the trace buckle.
HIP AND BACK STRAPS—1 inch to buckle in the pad and 1-inch hip strap sewed in a Cooper brace carrier with trace lugs to buckle on. Folded crupper to buckle.
TRACES—1½ or 1¾-inch double and stitched trace, 6 feet long, with clip cockeye riveted on; a very strong, well made trace.
TRIMMINGS—Full XC white metal trimmed.
Weight of harness, boxed for shipment, about 70 pounds.

BREECHING $5.50 EXTRA

No. 10K849 Price of harness, 1½-inch traces, without collar **$20.99**
No. 10K853 Price of harness, 1¾-inch traces, without collars **$21.99**
Add extra for breeching to buckle on the hip strap with extra strap on the breeching to snap in the trace carrier, making a double hip strap breeching. Price, per set, extra **$5.50**
Add extra for two ¾-inch tie straps with the harness .... **.52**
Add extra for single hip breeching to buckle on hip strap **3.18**
This illustration shows the double hip strap breeching furnished with this harness to buckle on the hip strap, and the extra hip strap to snap in the trace carrier, for $5.50. Add this price, $5.50, to the price of the harness if you want the breeching with the harness.
For price on collars see collar page.

QUALITY is remembered a long time after the price is forgotten. Our customers would rather pay $1.00 or $2.00 more for a good set of harness or saddle, strap work, collars, or fly nets, and get the best value that can be bought, than to pay $1.00 or $2.00 less and get an inferior article of poor quality. Quality considered, our prices are the lowest in the world.

Credit: Sears, Roebuck & Co.

FINE FARM HARNESS. Page from the 1908 Sears, Roebuck mail order catalog.

THE HARNESS AND TRAPPINGS

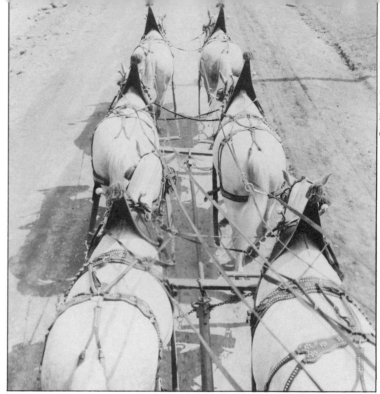

business leaving what was left to a very few.

In 1935 the Milwaukee Saddlery Co. claimed it had its best business in 10 years. There were still 17 million horses in the country and those companies able to stick it out did good business. This company said it regularly shipped 10 to 15 tons of harness in one day.

What might be considered as the last gasp was the fact that in 1955 for the first time in over 70 years of publication Montgomery Ward omitted harness for work horses from their farm mail order catalog.

SIX-HORSE TEAM. The wheelers were hitched to the wagon. The swing team's doubletrees were hooked to the wagon pole. A body pole between the swing horses was the connecting link for the leaders. Their doubletrees were hooked to the end of the body pole.

DRIVER'S-EYE VIEW. View of harnessed draft team from the driver's seat.

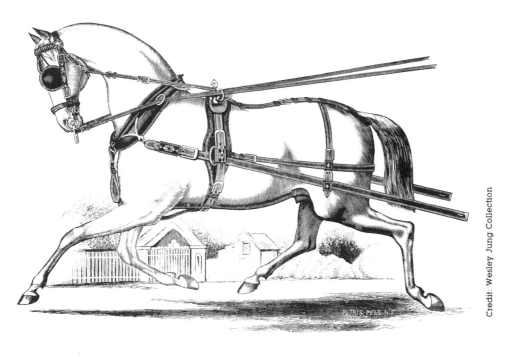

VARIOUS
HARNESS STYLES.
In Moseman's
Catalog of 1892
this was listed as
a "heavy Cab
Harness".

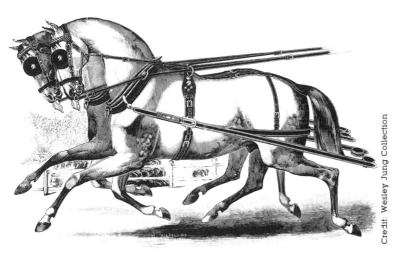

LIVERY USE. For country or city this was the "medium heavy Coach harness".

David Bradley Defiance Harness

Guaranteed for 5 years

Defiance
$59.90 Cash
1½-inch Traces
$6 Down

See inside
back cover
for Easy
Payment Ter

DETAILED ILLUSTRATIONS. Sears' 1942 catalog used very dramatic drawings to show their harness.

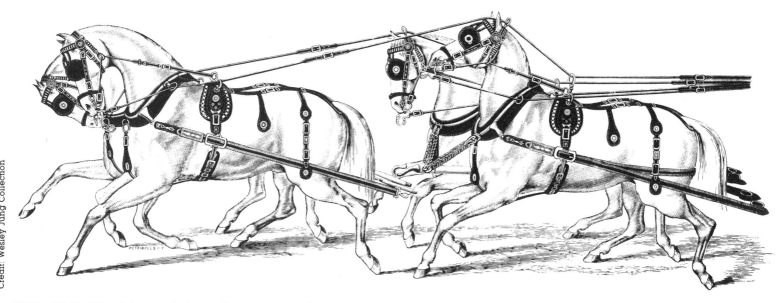

FOUR-IN-HAND. This coach team harness was "equipped with the latest patterns in trimmings, pad housings, monograms, crests, fronts and bits."

Black Beauty Breechingless

A Breechingless Harness That Really Fits and
Stays Where it Belongs

Full WALSH Ten Year Guarantee   30 Days Trial

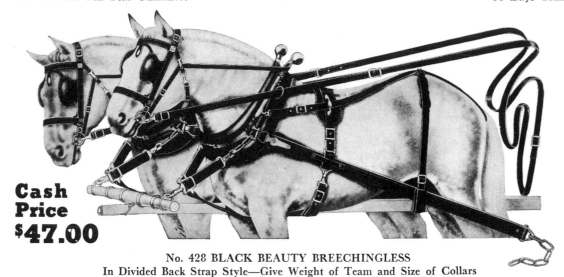

**Cash
Price
$47.00**

No. 428 BLACK BEAUTY BREECHINGLESS
In Divided Back Strap Style—Give Weight of Team and Size of Collars

Specifications the same as for No. 400 Black Beauty, except no breeching; 1⅛″ divided back straps, 1⅛″ hip straps; regular Walsh cruppers; same Fine Quality leather, hardware, and workmanship.

Per Set for Two Horses, Less Collars	Cash	Time
No. 428 —1½″ Traces, 1⅛″x18′ Lines..............$47.00		$51.50
No. 428XX—1¾″ Traces, 1⅛″x18′ Lines............... 49.25		53.75

Not Prepaid—F. O. B. Milwaukee

No. 426 Black Beauty Breechingless

Single Back Strap Style

Same as No. 428, except SINGLE BACK STRAP to back pad instead of divided to hames. Both back and hip straps are 1⅛″. Bridles with long checks to pads. Pads with check hooks and terrets. Regular Walsh cruppers.

Per Set for Two Horses, Less Collars	Cash	Time
No. 426 —1½″ Traces, 1⅛″x18′ Lines..............$46.00		$50.50
No. 426XX—1¾″ Traces, 1⅛″x18′ Lines............... 48.25		52.75

Not Prepaid—F. O. B. Milwaukee

Walsh "Yankee" or Hip Breeching

Of Finest Quality Heavy Harness Leather, to match either our Black or Russet Breechingless Harness.

No. YB110—2½″x24″ body, 1⅛″ side straps................$4.00 Per Set
No. YB120—3″ x42″ body, 1¼″ side straps................. 6.00 Per Set

PRICES FOR EXTRA EQUIPMENT

SWIVEL DEE HEEL CHAINS, add 40c per set.
BRIDLES with divided face piece, no extra charge.
BRIDLES, less blinds instead of regular, no charge.
BRIDLES furnished with long check rein with check-up straps to rump, instead of regular, add $1.50.
JOINTED BITS or JOINTED WIRE BITS, no charge.
ORCHARD HAMES instead of regular, no charge.
PLAIN WOOD HAMES instead of regular, add 85c.
For LYNITE ALUMINUM HAMES instead of regular, add $2.00.

HEAVY WOOD HAMES with brass tops instead of regular, add $2.25.
LOOP MARTINGALES with collar strap, add 85c per set.
For 20 ft. Lines on Nos. 300, 326, 400, 425, 426, and 428, add 75c.
WRAPPED END TRACES on No. 400, 428, and 426 Harness, add $1.50.
3″ SINGLE STRAP BREECHING instead of regular, add $1.00.
Brass Trimming, add $3.00 per set.

WALSH HARNESS COMPANY
245 EAST KEEFE AVENUE   »   MILWAUKEE, WISCONSIN

Credit: Author's Collection

Page Six

HARNESS THAT "REALLY FITS." 1939 Walsh Harness Co. catalog page.

EVERYTHING IN ITS PLACE. In this private stable of a wealthy gentleman, the harness room was always immaculate. (C-1896)

ORDERLY TACK. A good harness room had a place for every piece of equipment.

Credit: Wesley Jung Collection

Credit: Author's Collection

RAINWEAR FOR HORSES. Many catalogs offering horse trappings showed blankets for rain and cold weather.

PEST PROTECTION. The fly net was a widely used trapping. It kept flies off areas of the body that could not be reached with tail or mouth.

Credit: Gene Baxter Collection

FLY HOODS. Cloth headpieces, while not attractive on the horses, did control the pestering flies. The folding roof over the driver could be pulled forward in case of rain, or if the sun got too hot.

STRAW BONNET. Sometimes a driver would invent his own sunshade with a homemade straw hat. "Molly" pulling a bakery wagon in Waterford, New York, wears her bonnet well.

FLY BONNET. In addition to a fly net, this horse is wearing a fancy bonnet to keep flies away from its ears.

TASSELED NET. Another design of fly net.

FANCY SUNSHADE. Moseman's catalog offered this sun bonnet.

CHAPTER XII

THE FARRIERS AND BLACKSMITHS

FARRIERS WERE men who were accomplished in shoeing horses. It was a scientific profession, since properly shoeing a horse could eliminate a fault and prevent a horse going lame, or even cure lameness in a horse not correctly shod.

The blacksmith shop was the farrier's place of business. In a general way, he was called a ``smith'' or ``Smithy'' and could do remarkable things with iron.

Every hamlet and village had at least one blacksmith shop. The bigger the town the more there were. These shops were communication centers where news was heard, told and spread. Rumors were whispered and oft times started.

SCIENTIFIC METHODS. Mr. Brown's sales pitch was neatly painted on his building.

BULLETIN BOARD DOORS. Mr. Ulrich advertises that he does blacksmithing, but on a special bracket he has a couple of horse shoes suspended to advertise the fact he is also a farrier. The left side door has a herald advertising a Republican Rally, while the righthand door has a flyer advertising trotting races and signs for Kendall's Spavin Cure and Mitchell's liniment.

HOOF CHECK. Drivers checked their horses' shoes periodically. Here one of the caulks was found to be missing. Note rubber heel pad.

TYPICAL FORGE. The smithy's left hand is pumping the bellows forcing air in under the fire. C-1910.

198

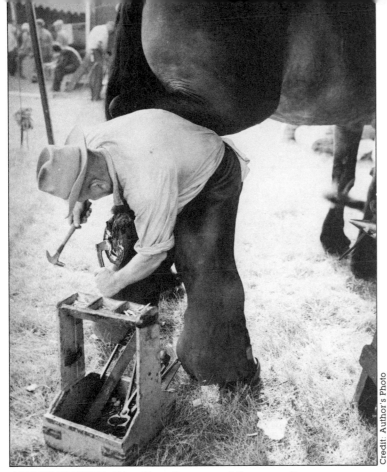

CIRCUS FARRIER. Ringling Bros. and Barnum & Bailey Circus carried their own farriers. Once the show was set up, horses requiring attention were brought to the blacksmith tent.

ANVIL CHORUS. The anvil was the farrier's work bench. It was generally set on a solid block of wood. Jesse Bain, who spent a lifetime shoeing horses and doing blacksmithing is shown here in his Midway, Kentucky, shop.

SHOEING TROT-TERS. An 1869 Currier & Ives print of a blacksmith shop. On the left are large hand-operated bellows that pump air below the fire in the forge. The small boy's job is to whisk flies away from the horse being shod. The gent with the grey beard is a typical patron; ears open for horse gossip, and always willing to pass on rumors he has heard.

"TROTTING CRACKS" AT THE FORGE.

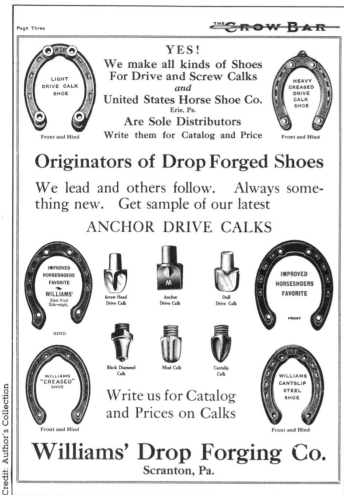
"WE LEAD AND OTHERS FOLLOW." This company advertised many styles of shoes in the November 1915 issue of *The Crow Bar* magazine.

HORSE SHOE NAILS. These specialized items came in many sizes as shown in this ad from the January 1910 issue of the *American Blacksmith* magazine.

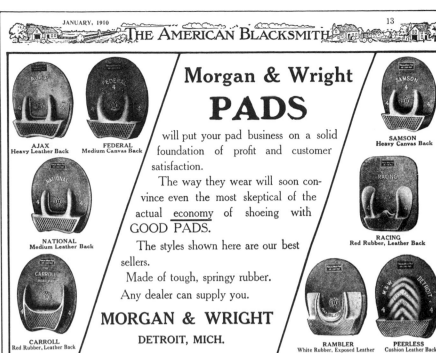

"Foot-Sure"
When He's Firestone Shod

No slipping, no hard pounding—just a steady, sure step and a comfortable, healthy hoof.

Firestone Hoof Pads are up to the widely known Firestone standard.

The quality means repeat sales.

Ask your jobber for prices

Firestone Tire & Rubber Company
Akron, Ohio—Branches and Dealers Everywhere

Firestone
HOOF PADS

Credit: Author's Collection

FIRESTONE SHOD. A 1915 advertisement in *The Blacksmith and Wheelwright* magazine.

ICE COVERED STREETS. These were a serious problem if a horse was not shod with caulks in shoes. Putting "chains" on horses seems rather novel.

KINDS OF SHOES.

Best Adapted for a Wide Variation in Shape of Hoofs.

Professor Lungwitz, the authority on scientific shoeing, gives the following as to the kinds of shoes that should be used for special purposes:

Shoe for a Regular Hoof.—Outer edge: moderately base-narrow, that is bevelled under, all around. Distribution and direction of nail holes are regular. The length is longer than the hoof by the thickness of the shoe.

Shoe for Acute-Angled Hoof. — Outer edge: strongly base-narrow around the toe, but gradually becoming perpendicular towards the ends of the branches. The punching to be regular except that the nail holes at the toe must incline inward rather more than usual. Length: rather longer than shoe above-mentioned.

Shoe for an Upright, or Stumpy, Hoof.—Outer edge: perpendicular at toe, but if hoof is very steep, base—wide at toe, that is bevelled downward and outward. Punching: last nail should be placed just beyond the middle of the shoe, and the holes should be perpendicular. Length of shoe not to be, at most $\frac{1}{8}$-inch longer than the hoof. In the case of a "bear-foot" the shoe must be long.

Shoe for Base—Wide Hoof.—Outer edge: the outer branch should be moderately base-narrow, that is bevelled downward. and inward, with the inner branch vertical. The punching upon the outer branch should have holes extended well back, while upon inner branch they should be crowded forward towards the toe. The length of this shoe depends upon the obliquity of the hoof.

Shoe for Base—Narrow Hoof.—Outer edge: the outer branch should be either perpendicular or base-wide, the inner branch strongly base-narrow. The nail holes in the outer branch should be crowded toward the toe, and, according to the circumstances, punched deeper than the wall is thick on account of the greater width of this branch. In the inner branch of the shoe the nail holes should be spread back to the quarter and punched light. The outer branch of the shoe should be one-fourth of an inch longer than the inner, and also wider.

Shoe for a Narrow Hoof.—Outer edge to be moderately bevelled under the foot at the toe, but elsewhere perpendicular. The distribution of nail holes should be regular, punched perpendicular, and towards the quarters inclining somewhat outward. About the toe the holes should incline inward. The length of this shoe depends upon the obliquity of the hoof—while no concaving is necessary.

A. Vineman & Weuerflien, Scarville, Iowa 1913

B. J. E. Morrison, South Carolina 1910

C. J. D. Norris, Drummond, Montana 1913

D. Matt E. Burdett, California 1913

E. Johnson & Bergquist, Mitchell, South Dakota 1912

F. Peter Pielmeir, Pennsylvania 1908

G. Potvin & Johnson, Great Falls, Montana 1914

H. J. W. Buettner, Norwich, North Dakota 1915

"KINDS OF SHOES" was a piece published in the August 1914 issue of The Blacksmith and Wheelwright magazine.

FORGES. Every farrier and blacksmith needed a forge. These models were advertised in the January 1910 issue of *The American Blacksmith*.

TYPICAL PRICES. 1914 horse shoeing prices in Brockton, Massachusetts. Prices are also listed for new steel tires on buggies and wagons.

Brockton boasts of being the only city in the world where blacksmiths and helpers work from 7 a. m. to 4 p. m., closing at noon Saturdays, with a total numbers of hours per week of 45.

Wages: Firemen, $19.00; floormen, $16.00 per week. Overtime, 42 cents per hour.

Prices in Brockton.

4 shoes	$1.75
2 shoes	.90
1 shoe	.45
Resetting	1.25
Sharpening 4 shoes	1.25
Toe weights	1.00
Side weights	1.00
Bar, 2 shoes	1.50
Hoof expanders	1.50
Leather pads	.40
Rubber pads and shoes, per pair	2.50

New Tires.

1⅛x¼	$7.00
1¼x¼	8.00
1⅜x¼	9.00
1½x⅜	10.00
1½x½	12.00
1⅝-1¾x½	14.00
1⅞-2x½	15.00
2-2¼x⅝	16.00

Credit: Author's Collection

ANOTHER CHAMPION.

A Nova Scotia Horseshoer Who Claims He Has Beaten Mr. White.

From the New Glasgow (N. S.) Enterprise.—In our issue of last week we had an article on the World's Record in Horseshoeing, said to be established by William White, a Scotchman in Vancouver, B. C., a short time ago. This article we copied from a Vancouver paper kindly sent us by some kind friend. This week we have received the following letter from our good old friend, Mr. George Campbell, of Avondale, who proves that he has beaten this record and can support it with affidavits and witnesses. Mr. Campbell's very interesting letter is as follows:

Dear Sir:—I read with interest in a recent issue of your paper an account of what is claimed to be a world's record in horseshoeing made recently by a Mr. White, of Vancouver, in which he claims to have forged the shoes from the bar and shod the horse complete in 24 minutes and 11½ seconds. It may be of some interest to know whether the fast time has ever been equalled by any other, although no official record has ever been made of the performance.

Without any effort at being egotistical, but as a plain statement of fact which can be supported by affidavits, if necessary, and witnesses to the performance still living, let me state: In the year 1870, I forged from the bar, with the assistance of a helper, to strike, twenty shoes, five complete sets, in one hour, or an average of one set every 12 minutes. About the same time I shod a horse for one Robert Robertson, of Piedmont, with No. 6 shoes—the watch being held by Mr. Donald R. McKenzie, of this place—in eight minutes, or forging and shoeing the above named horse in twenty minutes. I may add there were ten nails in each shoe. I have also, on a wager, turned and corked from the bar a set of horseshoes complete, only taking two heats to finish each—a record at that time that was unknown, and I doubt if it has ever been equalled.

I have no desire to gain fame by relating achievements of bygone days, but simply to show what has been done in the good old days of hand forging, or which stands equally by the performance of Mr. White.

Hoping you may find space for the above,

George Campbell.

Credit: Author's Collection

A CHAMPION HORSESHOER. This was in the October 1915 magazine *The Blacksmith and Wheelwright*.

The Old Blacksmith Shop

It Was a Man's World but the Sights, Sounds and Smells of Horseshoeing Appealed Strongly to Boys, Too

Roderick Turnbull, in the Kansas City Star

WHETHER in winter or summer, the blacksmith shop in the little country town was a gathering place for small boys. It was a good place for men to loaf, too, and carried a higher social status than the livery barn.

The livery barn somehow appealed to those hangers-on in town who liked a place to play cards or perhaps sneak a few drinks. But a blacksmith shop was a busy place. A man went there to get a job done, or perhaps visit with neighbors while the wife was exchanging her eggs for groceries. Mothers warned their sons to stay shy of the livery stable, but no such admonition was given concerning the blacksmith shop.

Fun for Boys

In Maple Run, my home town, my father was the blacksmith. The forge was not always busy on a winter day. On such occasions, it was a delight for us boys to play in the fire. Looking back, I recall there wasn't much around a blacksmith shop that a boy could harm. Maybe that is why we were allowed such freedom.

A boy's first lesson at the forge is to make a sizzler. This is a simple task that my father and his father before him must have taught dozens of boys. You just take a piece of iron, put it into the fire until it is red hot, then plunge it into the big wood tub of water at the forge's side. The water boils and sizzles until the iron is cool. You can do this all day.

Blacksmith's Apron Is Museum Piece

But it is more fun to pound the hot iron on the anvil, to make the anvil ring and the sparks fly. A piece of scrap bar iron can be flattened into a sword, or shaped into a dagger. If you get good, you can take a round bar and make it into a hoop, and if you are real good, you can heat the ends until they are white hot and weld them together.

A boy could spend a whole Saturday in the shop, even if he couldn't use the forge. For instance, it was nice to sit on a nail keg by the fire and listen to men talk while you sharpened your knife. We sharpened our knives on whetstones by the hour until they would shave the hair off our arms.

You could take the wood frames off old buggy tops and make shinny clubs out of them, and old wagon wheel spokes could be sawed into pucks to be used in shinny games when Mill creek froze over. The blacksmith shop, too, was the place to sharpen your skates, with files and a vise available.

When the blacksmith was too busy to watch, you could even take the clay pipes you bought for a nickel in Dave Stewart's grocery store and fill them with sawdust for a smoke. Truly this was a man's world.

The era of the blacksmith shop is over at Maple Hill. It ended when my father died, at near 81. The shop now is used by the Maple Hill township to house highway equipment and a fire truck.

The leather apron my father wore, his anvil, the old water barrel used for cooling hot iron, home made tongs and hammers and other paraphernalia of the shop now belong to the Kansas State Historical society in the state museum at Topeka. I wonder if boys of today can look at this stuff and imagine how much fun they are missing by belonging to an age that calls those things relics.

NOSTALGIA. In 1959 Roderick Turnbull reminisced in the Kansas City Star about "The Old Blacksmith Shop".

—Drawing by Einar Quist of The Journ

A boy could spend a whole Saturday just watching . . .

TIDY SHOP. White-washed walls give this blacksmith shop a neat and clean appearance. Racks hanging from the ceiling carry a large selection of horse shoes of many sizes.

SERVICE CALL. When a horse threw a shoe while on a city route, the farrier would be called out to handle the job. This photo is of one of 650 horses used by Tellings, Belle, Vernon Dairy in Cleveland in 1933.

CHAPTER XIII

THE FIRE HORSES

In THE 1860s horses began their careers of pulling fire fighting wagons, a career that would last 60 years more or less before they were shunted aside by the huge fire trucks.

In the beginning horses were stabled with their harness on their backs. Then in the early 1870s an ingenious fire fighter in St. Joseph, Missouri, devised a harness that could be suspended over the horse, and quickly lowered and locked in place once the horse had trotted from the stall to its place in front of the rig he was to pull.

Fire Department horses were kept in box stalls, or standing stalls facing out. They recognized their particular alarm signal and would be ready to dash out of their stall as soon as the barrier in front of them was automatically opened. Each horse knew precisely his spot and would trot out to it whether it was the center position on a pumper or ladder wagon, or the off spot on a hose cart. They were trained well and never faltered or missed their assigned location and rig.

These horses had to be surefooted—the pavement they galloped over could be slick with rain, snow or sleet. Galloping at full speed, they had to respond to the driver's signals on the reins avoiding other horse-drawn wagons, trolleys, people, and holes in the street.

Veterinary Dr. G.T. Unertl of the Milwaukee Fire Dept. said in an interview, "A fire

EYES ALERT, MANES FLARING. A team of galloping fire horses races down a Chicago street with Engine #4.

"FLYING" HORSES. A remarkable photo of Boston's Engine #18 in 1923. Only one hoof out of twelve is on the pavement at the instant the camera shutter clicked. Photo also shows the easy-on harnessing.

Credit: Boston Fire Dept.

Credit: Harold S. Walker Collection

JUMPING INTO ACTION. When the alarm bell sounded, there was a flurry of activity among men and horses. Ten, maybe fifteen, seconds later Engine #39 entered 67th Street in New York, horses galloping as they swung into the street. Seconds later the three greys, pulling Hook and Ladder #16, came out of the station. The matched horses are keyed up by the clanging bell behind them and the invigorating run.

THE FIRE HORSES

horse had to be active and run fast and pull. A nervous horse wasn't any good. He would not get any rest in his stall as he would be constantly anticipating the alarm.''

When an alarm went off signaling a certain company, the fireman on duty would throw an electrical switch that flung open the stall doors in front of the anticipating horses. The driver of each rig would be in his seat with seat belt snapped on. When his horses were in place he pulled a cord that released the harness down onto the backs of the horses. Men on the floor snapped the buckles.

At the same time, two horses came to the hose cart, three to the big ladder wagon, two to the chemical wagon, and one to the Chief's buggy. The entire operation took about 30 seconds.

In 1894 the Milwaukee Common Council sent a committee to find out the speed of the fire equipment. Their purpose was to check with stop watches the time between the alarm and when the equipment hit the street. The best time was registered by Engine Co. #4 on Second Street, which rolled out on the street in nine seconds flat. The three-horse steamer hit the street in 17 seconds.

Generally, the horses were hitched two or three abreast. The weight of the vehicle determined the number of horses. In the early years Milwaukee used three-horse spike teams, but later changed to three abreast. A ladder company in Lawrence, Massachusetts, tried a four abreast team.

In Rochester, New York, Ladder Company #1, an 85-foot Hayes aerial ladder wagon was pulled for a time by six grey horses hitched two abreast. This team, responding to an alarm, would have been a

MOVING OUT. This young team of Percheron geldings from Engine Company #32 moved its six-ton load down a Chicago street in 1908.

ANSWERING THE ALARM. 1908 scene in Brockton, Massachusetts. The driver of Engine #1 is intent on guiding his galloping team to the site of the fire.

truly spectacular sight when the six horses were at a full gallop.

Upon returning to the station after a fire run, all horses had their feet checked for loose shoes or lameness. Their nostrils and mouths were sponged out and they were given only a couple of swallows of water.

Winter in northern cities brought on some complications because of snow. Some rigs were equipped with runners. These were adaptable to hose carts, chemical wagons, and hook and ladder vehicles. Ottawa, Canada, did try runners on a steamer. Because it was slippery underfoot horses were held to a walk or trot.

E.T. Robbins, Associate Editor of the *Breeder's Gazette,* wrote a fascinating story about Chicago's fire horses in the November 15, 1911, issue. ''Draft Horses That Pull Fire Engines—Grade drafters are used and Chicago has 750 in the Fire Department. Those that pull the hose wagons weigh 1,200 pounds, and those on the pumpers and ladder wagons will go 1,400 pounds to 1,600 pounds and stand 16.2 to 17 hands.

''The big ladder wagons carry 40-foot long ladders. There is 25 to 30 feet between

TOUGH STEERING. Hook and Ladder Company dashes down an Albany, New York, street in 1911. These vehicles, carrying 40-foot-long ladders, maneuvered through traffic and around corners with the help of the very adept man on the tiller wheel.

THE FIRE HORSES

the front and rear axles. The light pumpers weigh 4-1/2 tons and have only two horses, but the heavier pumpers that weigh five to six tons require three horses.

"To fill the bill in the Chicago Fire Department, a horse must have an attractive, intelligent countenance, with a sprightly expression, medium sized head, wide between the eyes, straight faced, neat about the muzzle. The neck must be long and head well carried, for a short necked, low headed horse seldom travels out freely and willingly. A short back, well sprung ribs, close coupling, wide croup, muscular stifles and the best underpinnings are sought. Horses with long haired legs are avoided. Round feet and large at the hoof heads are sought.

"There is a great strain on the feet of these horses and it has been observed that dark hoofs are less brittle and more durable than white ones.

"Little training is given for fire work once a horse has been accepted. The new horse is given a little practice in coming out to his place under the harness when the door in front of his stall is thrown open. Most learn quickly—many learn in two or three days.

"The fire horses are fed four to six quarts of oats apiece, three times a day, and a bran mash about twice a week. Prairie hay is fed at night only and they get 12 pounds of hay per horse per day.

"Usually the new horses are between five and six year old geldings. They stay fit for fire work for three or four years on the average. Some horses last 10 years and some have seen 15 years service. They are discarded mostly for lameness from galloping over the hard paved streets.

HORSE GOES DOWN. An incredible happening in Utica, New York, in 1904 was recorded on film by an alert photographer. Unbelievably, the fallen horse scrambled to his feet and continued the run. The engine never stopped during this astonishing episode.

ESSENTIAL EQUIPMENT. A searchlight engine in New York City, C-1905.

Credit: H. J. McGinnis, M.D. Collection

ON RUNNERS. The Ottawa, Canada, Fire Department fitted out this LaFrance pumper with runners for winter work. This was rarely, if ever, done in the U.S.

Credit: Harold S. Walker Collection

Credit: Ron Ryder Collection

HOSE CART. 1910—Company #15 Milwaukee.

Credit: Author's Collection

"Fire horses have a vigorous constitution and are always in fine fettle. They are essentially a draft type, but each one handles himself with the ease of a light horse."

The following interesting squib appeared in the June 11, 1914, *Breeder's Gazette.* "A new hook and ladder truck recently added to the Chicago Fire Department started out of the station to answer an alarm but it reached the repair shop instead of the fire behind a pair of horses.

"Firemen have a life-long experience in arriving with horsedrawn vehicles and it, naturally, nettles them when the motor of the new fangled truck goes dead in time of greatest need.

"The gas engine cannot compete with the horse in reliability and adaptability."

In 1890 Milwaukee's Fire Chief Foley reported to the City Council, in his annual report that oats cost 38¢ a bushel, and hay $10.58 a ton. He said, "the department had 112 horses in service. 29 horses purchased in 1890 cost an average of $200.00. Three horses were killed, two responding to alarms, and one while in pasture."

Chief Foley recommended "an additional barn be provided for extra horses and used as a training school."

"Often when a new horse is placed in a fire engine house, that horse will not lie down for two or three weeks, and then goes down

KEEPING THE PUMPERS WORKING. The unsung horse heroes in the Fire Department were those who came to the conflagration with a load of coal for the pumpers, as seen in this photo.

Credit: State Historical Society of Wisconsin

THE METROPOLITAN. Built in 1900 by the American Fire Engine Co. of Cincinnati, this Metropolitan was originally used in Milwaukee and later by the Oconomowoc, Wisconsin Fire Department.

FRONT VIEW of the Metropolitan. Note the bell under the driver's foot rest, and the drag shoes on the pole, instead of drag wheels.

REAR VIEW of the Metropolitan.

TAKING A SHARP CORNER. Ladder Co. #1 pulls out of the station in Norwich, Connecticut. This is a Dederick Aerial Ladder that is 85 feet long when extended.

THREE GALLOPING GREYS. The team pulls the huge ladder wagon over the city streets to a fire. The distance between axles could be 30 feet on a wagon such as this, carrying 40-foot ladders. Rigs this length had a tillerman to handle the rear wheels on the turns.

THE FIRE HORSES

because of sheer exhaustion. When newly purchased, a horse is generally too fleshy to endure a quick and hard run. That factor can be obviated by having extra horses for training purposes. It requires fully one year in the service before a horse can become thoroughly accustomed to the service. Inexperienced horses should receive at least one month's training before being placed in active service. Then a horse should be placed in one of the fire engine houses on the outskirts, where the service is not as arduous. As the horse develops, his service should, accordingly, be increased by sending the horse to a company with a greater alarm schedule. The same holds good for horses assigned to the downtown companies. That is, after a while those horses should be placed in outskirts companies.

"I can justify claim for the horses of the Department that they are as fine as can be found in the nation. They are all, or at least 95% of them, Wisconsin bred. Such we have found to be the best, as there is no trouble acclimating them.

"A horse first of all is examined by the Department Veterinarian as to age, style, height, weight and blemishes. If found satisfactory a horse is generally hitched to a hose wagon alongside of an experienced horse and driven for a distance of a mile. Such a horse frequently comes back panting, which proves that the horse is not adapted to the service. If a new horse is not winded after such a trial, such a horse is perfectly safe for the service and should be purchased. With care and attention, such horses prove to be well fitted for the Department."

The Veterinary Surgeon, Charles H. Or-

TROY, NEW YORK. In this 1909 photo of the fire house at State & Third Street, the fire chief is driving an automobile—an omen of changes ahead.

213

THE FIRE HORSES

CANADIAN COMPANY. Montreal's watertower engine operating out of the central fire station.

mond of the Milwaukee Fire Department, commented in his January 1, 1897 report: "The health and condition of the horses is good. We have been fortunate in regard to accidents and loss by death. The following deaths occurred, none, however, through carelessness or ignorance on the part of man.

"June 18, 1896—Bay Horse #45 on Truck #1 dispatched on account of incurable disease.

"June 20, 1896—Cream colored horse on Engine #21 from intestinal calculi. This animal, during five years of service, had a great many attacks of intestinal cramps, but would recover from apparent death and take his place in the service as well as ever.

"August 23, 1896—Gray horse #139 on

HOSE EXTENSION. Water tower of Engine Co. 24 in Los Angeles.

TYPICAL FIRE HORSES. Chemical wagon on the Rensselaer, New York, fire department in the 1890s. This is an excellent view, not only of the harness, but of the type of horse used by fire departments.

SPECIALLY EQUIPPED. W. S. Nott Co. of Minneapolis had this Chief's buggy in its line of equipment.

AERIAL HOOK AND LADDER. This wagon, illustrated in the W. S. Nott Co. catalog, could be placed in the middle of a 40-foot street and the ladder leaned against a building without overturning.

Truck #6 of apoplexy while responding to a test trial on the Chemical Engine.

''November 5, 1896—Brown horse #121 on the Water Tower of gastric hernia.

''The new engine houses to be put in service will require 10 horses. I would suggest that the Department be equipped in every respect with first class animals this year, 1897, for the reason that the horse market is now very low and there is every reason to expect that in another year the market will not only be much higher, but that good animals will be hard to get at any price.

''The care of the horses has been exceptionally good. The men in charge are natural horsemen and have a love for and peculiar delight in treating the animals kindly and giving them the best possible care and they are rewarded by the horses giving them their best efforts in responding to the alarms.

''Most of the Company quarters are now equipped with swinging harnesses and I trust the old way in which the horses are obliged to sleep in their harness will be forever abolished.''

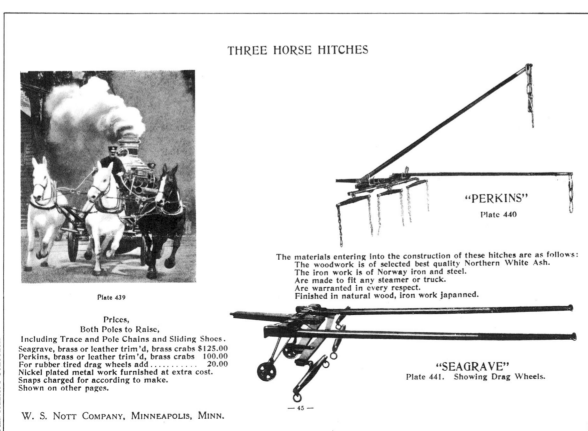

THREE HORSE HITCHES

Plate 439

Prices,
Both Poles to Raise,
Including Trace and Pole Chains and Sliding Shoes.
Seagrave, brass or leather trim'd, brass crabs $125.00
Perkins, brass or leather trim'd, brass crabs 100.00
For rubber tired drag wheels add 20.00
Nickel plated metal work furnished at extra cost.
Snaps charged for according to make.
Shown on other pages.

W. S. NOTT COMPANY, MINNEAPOLIS, MINN.

The materials entering into the construction of these hitches are as follows:
The woodwork is of selected best quality Northern White Ash.
The iron work is of Norway iron and steel.
Are made to fit any steamer or truck.
Are warranted in every respect.
Finished in natural wood, iron work japanned.

"PERKINS"
Plate 440

"SEAGRAVE"
Plate 441. Showing Drag Wheels.

— 45 —

BREAKAWAY HITCHES. Page from a catalog of a fire apparatus manufacturer, W. S. Nott of Minneapolis. Note the drag wheels on the pole. When a pumper arrived at a fire, a pin was pulled from the pole and the horses were walked down the street and away from the fire. The drag wheels would keep the rigging from skidding on the ground.

215

CHEMICAL COMPANY #7 MILWAUKEE, 1910. Note two chemical tanks in the body of the wagon.

PRIDE IS SHOWING. Firemen were proud of their horses. This group poses in front of a Milwaukee Fire Station in 1910. The three horses on the left pulled Engine Company #18 Steamer.

FIRE "SLEIGH." Ottawa, Ontario, as many northern cities, resorted to sleigh runners on its equipment during the winter months. This is Hose Co. #8.

UNIQUE HITCH. The Rochester, New York, Fire Department utilized horse power in a very novel way. This could possibly be the only department that strung out three teams in this fashion. The horses are hitched to a heavy Hayes Aerial Ladder, which when extended was 85 feet long.

SIDE-BY-SIDE. Probably the most unusual hitch of four fire horses was this team on the Lawrence, Massachusetts Fire Dept. Originally the Department used three horses in the summer on this equipment, and added two out front in the winter. Later the Department settled on four abreast to pull the Kress wagon that carried 450 feet of ladders.

Credit: H. J. McGinnis, M.D. Collection

216

UNICORN HITCH. A hitch of three horses was not commonly used. The Hook & Ladder Co. #7 operated in Minneapolis.

SUSPENDED HARNESS. The Bay View Fire House on South Kinnickinnic Ave., near Otjen Street in Milwaukee, C-1910. This is an unusually good photo of the harness as it was suspended over the spot where the horses would be seconds after an alarm sounded.

FOUR-HORSE RIG. The Minneapolis Fire Dept. used a 4-horse team—two abreast on Hook & Ladder Co. #11.

EASY-ON FOR FAST STARTS. Engine Company #11 in Troy, New York. The collars on fire horses were clamped at the bottom. Other horse collars were buckled at the top.

UNUSUAL ONE-HORSE RIG. The chemical cart had an 80-gallon capacity tank and carried 300 feet of 1" diameter rubber hose. Chicago Fire. Dept.

FIRE HORSE FUNERAL. In memory of a faithful Fire Dept. horse, the sign on the wagon says, "Old Pat has gone to his last fire".

CHAPTER XIV

ADVERTISING WITH HORSES

THE USE of six- and eight-horse exhibition teams to advertise companies and their products seemed to be a natural adjunct to the horse business. In the horse-drawn era, everybody loved the sight of well-groomed and flashily harnessed horses.

Exhibition teams showed up in parades on special occasions in towns and cities. They performed their figure eights and other magnificent maneuvers at a trot at the county and state fairs. They were a superb attraction in those days. (And, indeed, an astounding number of companies, breeders and draft horse owners continue the practice today.)

One of the reasons for the great success of the circus street parades was the splendid teams of six, eight and more matched horses pulling the bandwagons, cages and calliopes.

In a way, these exhibition teams promoted draft horses even more than they promoted products or companies.

In the May 4, 1910 issue of the *Breeder's Gazette* a letter appeared from Taylor & Jones, horse dealers of Williamsville, Illinois, ''The six-horse teams seen at our great shows are certainly worthy of appreciation of all horse lovers.

''The men who take such pride and ex-

HIGH-STEPPING CHAMPIONS. The Armour and Company had a beautiful team of dapple grey Percherons that were champions in the U.S., as well as England in the 1906-1908 era. Billy Wales was the driver. The wheelers were Big Jim and Henry; the swing team were named Abe and Phil; while the leaders were Dude and Mack. This team in 1908 was a featured attraction on the Sells-Floto Circus.

218

PABST PRIDE. The Pabst Brewing Co. team was the pride of Col. Fred Pabst, who bred Percherons on his Oconomowoc, Wisconsin, farm. C-1904.

ADVERTISING WITH HORSES

pense to secure the very highest type at extremely high prices should be appreciated by the world at large. We think one of the best advertisements that could possibly be given to the public is to see such specimens dressed in the most expensive harness and hitched to wagons, regardless of cost with everything done that is possible to make them appear pleasing and wonderful to the eyes of men, women and children.

"We think they have done a great deal to advance the interests not only of farmers but of the importers as well.

"To see a driver handle six horses, seated in his high position, turn them around on 20 feet of driveway, stopping at any angle, then dashing away at lightning speed is a sight.

"We remind each farmer and breeder and stockman as he admires the beautiful turnouts in six-horse teams that each of these horses has been bred and raised by some practical fellow farmer or stockman.

"We think the exhibitors of the six-horse teams have done more for draft horse interests at large than all other advertisements, and we certainly thank and appreciate men who will take the pains and spare no amount of money to find, buy, groom and educate these horses for public exhibition."

In the same issue the *Gazette* printed another letter from a horse dealer in Ottawa, Illinois, W.E. Prichard, who commented, "These exhibitions are instrumental in setting high ideals and standards for breeders, farmers and team users.

"One of the valuable lessons is that it takes time to make a show team. Simply making a horse fat does not make a show horse. A typical show horse must be so handled while he is being conditioned and mannered as to enable him not only to retain the action he has by nature, but to develop ac-

BORDEN STYLE. Borden's greys were a stylish-looking team.

tion and style at the same time.

"I believe that nothing has ever done so much to create a desire among farmers to improve their draft horses as the exhibition of the great six-horse teams at the Fairs.

"A word of caution to the breeder—remember that simply a big horse will not do. While he must stand up he must in no sense be leggy. He should have width and sub-

Credit: Anheuser-Busch Co.

CLASSY CLYDESDALES. This is the Anheuser-Busch Brewery team of the 1930s. Many brewers used big teams for advertising purposes.

BEAUTIFUL BEL-
GIANS. Peffer Fur-
niture Co. of Stock-
ton, California,
made the fairs in
the west with this
team of Strawberry
Roan Belgians.

ADVERTISING WITH HORSES

stance enough to make him well proportioned.

"His head must be well formed, indicating intelligence. His neck must be of good length, coming up nicely out of the top of his high withers. He must have good sense and plenty of spirit, but no frothy disposition will do. It goes without saying his feet and legs must be faultless. You say, 'the standard is high'? Yes, so it should be and yet it is not unattainable."

Two weeks later another letter appeared from McLauglin Bros., a horse importer of Columbus, Ohio: "It is an undisputable fact that nothing has advertised the usefulness of draft horses in harness more than the exhibition of six-horse teams at livestock shows.

"These teams not only advertise the draft horse, but they educate the general public and cause the demand to increase for horses of superior quality. We appreciate the prominence given draft horses by those who exhibit teams of high class geldings and we hope they will continue to make these exhibitions and still further promote this greatest agriculture industry."

Then in the June 1st issue the Gazette printed a letter from one of the owners of a famous six-horse hitch which gave his interesting views. The letter was signed by Mr. Fred Pabst, whose brewery was in Milwaukee, but the message came from his stock farm near Oconomowoc, Wisconsin.

"Showing this high class stock to farmers

and breeders at our great exhibitions gives them an opportunity to form ideals on practical lines. It should greatly influence the gradual but positive improvement of our draft stock.

"To packers and brewers the showing of six-horse teams is strictly a business proposition, an advertising medium. They spend thousands of dollars yearly in printer's ink, and have found through years of experience that is a paying investment.

"But leaving aside the actual advertising value that the packers and brewers receive, they are great benefactors to the agriculture interests by influencing breeders to raise higher class stock."

GLEAMING RIG. Wilson & Company, Chicago, meat packers, made many fairs and stock shows in the 1930s with this team of Clydesdales.

Credit: Wilson & Co.

DAPPLED BEAUTIES. In 1909 Swift & Co. was showing this classy-looking team.

CHAPTER XV

FARMING WITH HORSES

WHEN all is said and done, the use of horses to do all the farm work did not cease because the horses were not dependable, reliable and adaptable. Horses lost out because tractors could do the various jobs faster, thus more economically.

In the article, ''Choosing Farm Horses'' in the January 22, 1914, issue of *Breeder's Gazette,* farmers were gently scolded. These are excerpts:

''Most farms are overhorsed—so many horses are not quite equal to a full day's work each day. Some horses are old and weak.

''A weeding out system is in order. The usual excuse for retaining undesirable horses is that they will not bring much money if sold; even though that is true, one cannot afford to keep an inefficient horse.

''The poor farmer keeps poor horses and the rich farmer keeps good horses.

''Let every farmer remember these facts as he sorts out the horses to be retained for another year's work, or buys additions for his teams.''

An editorial in the October 14, 1915, issue of the *Breeder's Gazette* commented, ''We are at the beginning of a momentous, though

Credit: Author's Photo

MAKING HAY. Team of Belgians mowing alfalfa in Wisconsin.

ON THE WAY TO WORK. Percheron mares are driven to the field.

FARMING WITH HORSES

gradual change in the relation of power to farming. No thoughtful man can hold with sentimentalists that machinery is not going to greatly decrease the use of horses in farming operations.

"It is no use arguing the point—power farming will come because it will justify itself in economy and efficiency."

In 1921 the *Gazette* compared horse populations for the past decade:

1910—Horses on Farms	19,833,000
Mules on Farms	4,210,000
1920—Horses on Farms	20,142,455
Mules on Farms	5,450,623

World War I had created a temporary reprieve on the coming change.

"There can be no question that motor trucks and farm tractors are providing a boom and a blessing to many farmers, but it becomes more and more apparent that they are not replacing work horses so quickly and so seriously as some would have us think," the *Gazette* said.

Many factors entered into decisions on making the switch. The long haul to market

PLANTING TIME. Four-horse grain drill in operation near Halstad, Minnesota.

WELL-MUSCLED, carefully groomed and in good condition, these 18 Percherons are ready for a long, hard day's wo

was a big factor for many farmers. Others who were short on land looked at facts listed in the farm magazines. If a farmer owned 240 acres he had to set aside 30 acres to maintain 10 head of work horses and another 30 acres to maintain 12 growing colts. Tractors did not require this acreage, and they did not need to be fueled daily, whether they worked or not.

There were those who were quick to rebut that tractors have a high first cost and high maintenance. Plus they damage the land, they supply no fertilizer and, unlike horses, do not replenish themselves.

And as one paper said, ''A three-year old colt is a lot younger than a three-year old touring car.''

WEEDING THE CORN. Each of these teams could work fifteen acres a day pulling two-row cultivators.

orses such as these were dependable, reliable and adaptable sources of power on America's farms.

McCORMICK REAPER. Cutting and binding the grain.

HAYSTACKS. Haying in Wisconsin, horses pulled the rake and hayracks.

HORSEPOWER. Eighteen Percherons such as these could do a lot of farming for any man.

WASHINGTON WHEAT FIELD. This state produced 42 million bushels of wheat in 1920, harvested with outfits like this—in this photo, 165 horses pull five grain harvesters.

SUMMER SCENE. Threshing grain near Sioux Falls, South Dakota. Two teams, with big loads, are on far side of the thresher. The wagon load of shocks on this side of machine is almost unloaded. The grain wagons are on the left in this photo.

PREPARING A FIELD. Eight horses pull a disk and harrow.

MULE POWER. Twenty-seven mules pull this combine near Walla Walla, Washington.

ONE DOLLAR A YEAR
CANADA, $2;   FOREIGN, $3.50

THE BREEDER'S GAZETTE

Established 1881 "The Farmer's Greatest Paper" $1.00 per Year

September 3
1914

Sanders Publishing Company, Chicago U.S.A.

MAKING HAY. Filling the hayloft is well illustrated on this *Breeder's Gazette* cover. It was also horse power that pulled the huge fork full of hay from the loaded wagon up into the mow.

TILLING THE SOIL. Two views of fall plowing with reliable power.

DEMANDING WORK. Near Dell Rapids, South Dakota, three horses pull a corn binder. Another team and wagon accepts the corn stalks. Harvesting and fall plowing were considered the hardest work on the farm for horses.

DISKING A FIELD. These two 4-horse teams were working on the Chicago Tribune's experimental farm.

CHORE TIME. Horses were handy for many jobs. Here, two Belgians move a hog waterer.

SPREADING MANURE was a chore that had to be done day in and day out.

BRINGING HOME THE CORN. A load of ear corn comes in from the field in Indiana.

STORING THE HARVEST. Horse-powered grain elevator was practical. As the load of corn was up-ended, the ears slid into the elevator and were carried up into the corn crib.

MULTIPLE HITCH EQUIPMENT

In the few years since the big hitches were introduced into the central west, they have saved millions of dollars to the farmers who were using them.

In addition to saving man power and using horse power more efficiently, they help solve the farm relief problem. Horses and mules eat the products the farmer himself raises, so he need not pay out hard-earned dollars to keep his mechanical power running.

This increased interest in horses and mules is also shown in many places where farmers are clubbing together, purchasing a stallion and raising their own horses. Many farmers consider the good, heavy, well-bred horses they are raising one of their best investments which will bring them profits for the next fifteen or twenty years.

Every dealer should therefore stock a complete set of big hitch equipment and lose no opportunity of showing his customers how to make more money through the use of big hitches.

FIVE HORSE HITCH—3 and 2

This is the most popular of all five horse hitches. Each horse has plenty of room and is in plain view of the driver.

The horses are handled very easily on the turns and because of their freedom of movement, the horses walk along rapidly and can exert their maximum pulling strength.

Every farmer who has five horses should use this hitch. He will find that he can handle his farm work without hiring an extra man. See price list for equipment needed, cost to you, and suggested resale price.

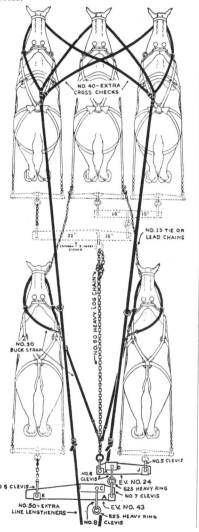

FOUR HORSE HITCH—2 and 2

This method of driving four horses is rapidly replacing the old four abreast hitch. Tests made by agricultural colleges have shown that four horses develop from 25 to 30% more power, when hitched in tandem or two and two. It is surprisingly easy to handle on the turns, the horses walk along rapidly and each horse can exert its maximum strength.

A great many farmers try this hitch first, find it successful, then come back and buy more equipment, and start using six or eight horses. See price list for equipment needed, cost to you, and suggested resale price.

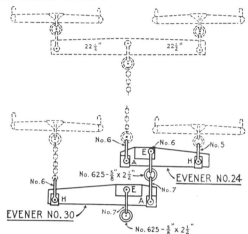

SIX HORSE HITCH—3 and 3

A very practical and all-around hitch. The farmer with six horses will find that using his six horses in this way will be a very profitable and simple way of working his farm.

Six horses will easily pull a 14 inch gang plow with a harrow behind. They have plenty of power to handle a 10 foot binder, small combine, a corn harvester, with or without motor or a small tandem disc.

See price list for equipment needed, cost to you, and suggested resale price.

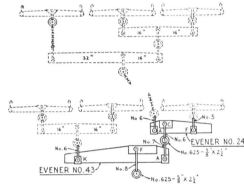

EIGHT HORSE HITCH—4 and 4

As farmers become more familiar with big team use, they will demand larger hitches. Many of them are now using this eight-horse hitch and in the states where they have more horses on each farm, some are already using ten and twelve horses and driving them with one pair of lines.

This is the most popular 8 horse hitch. It has plenty of power for a 10 foot tandem disc, 3 bottom plow with harrow on behind, and a combine under any conditions.

See price list for equipment needed, cost to you, and suggested resale price.

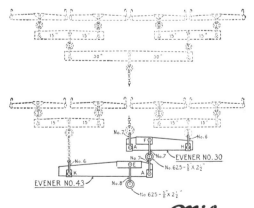

MONARCH—The Monarch of All America's Work Harness

Milsco

COMBINATIONS. The Milwaukee Saddlery Co., a large manufacturer of work horse harness, published these diagrams for multiple-horse hitching in their catalog.

MILK CANS. Farmer's rig for taking his daily milk production to the local creamery near Troy, New York C-1910.

GROWING HORSEPOWER. Sturdy Belgian foals enjoy the early summer sun on an Illinois farm.

FRIENDS. Farm draft horses were gentle and easygoing.

RENEWABLE RESOURCE. One of the advantages of farming with horses was raising your own replacements. This scene was photographed in Illinois.

HORSE PULLING CONTEST

HORSE PULLING CLASSES AND PRIZES

ENTER YOUR HORSES FOR SHOW — IT WILL

MAKE YOU EXTRA MONEY!

Each contestant must purchase exhibitor's ticket.

Class A—
Teams weighing under 3,000 pounds.

Class B—
Teams weighing 3,000 pounds or over.

Prizes for each class based on tractive pull.
1st: $25, 2nd, $15; 3rd, $10.

A tractive pull of 2,000 pounds equals:
(1) Lifting 2,000 pounds straight up out of a well.
(2) Pulling four 14-inch plows 6 inches deep in clay loam.
(3) Starting a wagon load of 20,000 pounds on concrete pavement.
(4) Ten horse power (H. P.) when extended for 27½ feet in 10 seconds.

For details write or see the county agricultural agent.

RULES AND CONDITIONS

No whipping. The teams are to be weighed in the presence of at least one of the judges the hour before the contest, and to appear on the grounds with the certificate of weight at the hour of the contest.

Neither horse shall be given on advantage at the evener. It is best that the team be shod. A tough grass sod will be secured for the pulling site if possible. The spectators are asked to stay back of the ropes and to refrain from cheering while the team is pulling.

If two teams fail to pull the last load 27½ feet, the award is based on the distance each does pull it, if any. If they fail to move the last load at all, the award is based on the number of seconds it took each to make its last successful pull, the faster team winning.

Credit: Author's Collection

CONTEST RULES. Rules and conditions of a 1936 horse pulling contest.

POWERFUL TEAM. In 1934 a Sandwich, Illinois team set a new record for Illinois teams by pulling 3,400 lbs. on the dynamometer.

POWERHOUSE OF A TEAM. Rock and Tom, weighing 4,350 lbs., pulled 3,900 lbs. on the dynamometer. They were from Pima, Ohio. That's the equivalent of a rolling load of 130 tons—and they pulled it 27-1/2 feet.

CHAPTER XVI

CIRCUS DAY

WHEN the circus came to town—and a circus did come to every village, town and city—it seemed a veritable horse fair had arrived. A big circus that traveled by railroad might have had 200 dapple grey Percherons working as four-, six- and eight-horse teams and another 200 performing horses.

Circus day was one of excitement, wonderment and absolute pleasure. Only those who could afford the 50¢ ticket went to the performance under the big tent. But those without the half a buck still had a free show to watch—the huge red wagons had to be unloaded from the flat cars, then hauled to the circus lot. Elephants, camels and zebras moved through the streets, as did cage wagon after cage wagon of wild animals that roared, snarled and coughed behind panel sides. What was in a den kept a kid guessing and imagining.

Part of the free show was watching the long string teams hauling these cages and the big red baggage wagons from the railroad yards over the streets to the showgrounds.

There was even more action when these

CIRCUS PARADE. Once the circus tents were set up and ready for the performance, the street parade left the showgrounds and headed downtown. The purpose was to let the townspeople know that today was ''Circus Day'', for sure. This advertising stunt, usually staged around 11:00 a.m., was about the last work the draft horses had to do until late afternoon, or early evening. In this photo, people of Ogdensburg, New York, jam the streets to see and hear and smell the exciting event.

CIRCUS DAY

horses spotted the wagons in their proper places on the lot. If it was a rainy day and the ground was soft, the massive wagons sank in the mud. Then it was doubly exciting to watch the powerful Percherons work.

When the tents were up and all was set, a bugle sounded the parade call. And out onto the street came the most wondrous spectacle ever seen by the townspeople—the circus street parade! It went right downtown, circled around then came back to the showgrounds.

The free street parade announced that "Today Is Circus Day"—no mistake. It was a spectacular form of advertising. Horses, horses and more horses never seemed to stop clip-clopping by, and the towners appreciated good-looking horses.

One of the Ringling brothers, Alf T., wrote in the 1895 route book of the circus, "A circus without horses would be like a kite without a tail. The noble horse is an adjunct as absolutely necessary to the show as the rudder is to a ship. It is indeed doubtful if there would have ever been a circus had there never been a horse. With the Circus the horse is a necessity in every department. The most important of all the horses with the show are the ring stock or bareback horses. A way might be devised for transporting the show from the railroad cars to the showgrounds. Horseless vehicles might be contrived to do this work in a horseless age. A parade might be made without the horse and could be dispensed with in many aspects where he seems, by habit and custom, indispensible. But no contrivance run by steam, electricity or naphtha could be invented to take the place of the live white horse prancing around the circus ring with a rider on its back. The old-time cry of the clown 'bring out another horse' could hardly be paraphrased into

WET WORK. On a rainswept morning the men and horses carry on, but city folk are scarce.

CIRCUS COMES TO TOWN. Towners crowd around the stock cars as the "baggage stock" unloads. This is the circus term for draft horses.

BARNUM & BAILEY. Two huge wagons arrive at the lot in Milwaukee in 1909. Another team of six horses is moving in. The big tent will be erected in the open area on the right. Scores of towners line the road and mill around the lot, not wanting to miss anything.

HEAVY HAULING. The light plant wagon carries a massive load, as it moves onto the lot on the Al G. Barnes-Sells-Floto Circus in 1938. The wheels are 8'' wide and the tires are 1'' thick.

REINFORCEMENTS. The team of six blacks is going to get some help from the six greys in moving an extra-heavy wagon for the Cole Bros. Circus. Notice all the townfolk in the background—circus day intrigued them.

ON THE WAY TO SET UP. Once the wagons are on the street, teams of horses take them to the showgrounds. People living along the route have a ringside seat to watch all the activity. In this photo a stake driver is moved to the lot.

'bring out another motorcycle,' with any degree of satisfaction, even if the rider could do his tricks upon some such horseless horse.''

"Billboard" magazine in its Sept. 30, 1905 issue reviewed the Barnum & Bailey Greatest Show On Earth in the far West.

Here is an excerpt: "A new feature was introduced at San Francisco, which added immensely to the program. This act was a parade of 260 of the show's draft horses. The first circus horse show ever given was presented to the San Francisco public. It made one of the greatest hits that can be imagined. As the line of magnificent animals wound around the track, four abreast arranged in quads according to their color, blacks, browns, sorrels, roans, dapples, greys and whites, a murmur of admiration arose through the audience and they received a hearty round of applause. Many wondering remarks were passed as to how the Show could keep them in such good condition and have so much work for them to do. At every performance following the introduction of this new feature, the parade was given, and from the applause that always followed the entree of this equine pageant, it could be easily seen that Frisco possessed its quota of lovers of fine horse flesh.''

Circuses seemed to like the dapple grey Percheron as the basic draft horse.

Mr. E.T. Robbins, Associate Editor of the *Breeder's Gazette* wrote a story for the August 19, 1915 issue on grey horses. In it he said, "The Ringling Bros. buy greys because they are known for their vigor and endure hot weather better than those of other colors.

"The Ringlings also say that grey horses as a class are most intelligent, most tractable and the least excitable.''

Another one of the Ringling brothers, John, wrote in the September 1919 issue of the *American Magazine* a fascinating story about how he and his brothers ran their circus. Here are excerpts of his comments:

"As to horses, there exists in the human mind a love and admiration for the horse that is almost beyond comprehension. There is

CIRCUS DAY

some psychic connection between the human and the equine animals, which is perhaps inherited...

"It is largely because of this universal love of horses that Ringling Brothers have opposed the idea of motorizing the circus. Looking further, we know that, within a short time, the horse will be almost as much of a curiosity to the public as the giraffe was a generation ago. Do you realize how few of the new generation ever have seen a four- or six-horse team? Our circus horses, which, of course, are selected with the greatest care, are almost as much of an attraction, either in a city, or in the country, as are the rare animals collected from all over the world to form the educational feature of the shows.

"Besides that, we do not believe that people want to see machinery in motion when they come to a circus..."

In the days when horses ruled supreme, the circuses seemed to respond by making the horse the cornerstone or even the foundation of any circus performance.

To depict the performing horses at their beautiful best, a number of circus posters are shown in this chapter. These illustrate the remarkable trained animals that drew oohs and aahs and applause from audiences in small towns and big cities alike.

"HOOK ROPING." When a team of eight horses cannot move a heavy wagon over the soft ground, the Boss Hostler signals in other teams. A hook on the end of the chain is placed on a steel ring on the corner of the wagon. Thus, with th extra horses, probably 24 in the photo, the wagon is spotted where its load will be needed.

MUDDY LOT. The circus moved, rain or shine. A day of rain only made it miserable for drivers and horses.

IMPRESSIVE SIGHT. Twenty-four dapple grey Percherons pull this bandwagon, not because it is so heavy, but because they present a great spectacle on the city streets.

CANVAS TROUGHS. The circus was always self sufficient. Here they set up their own portable canvas water troughs. Cool water is pumped in from a nearby fire hydrant.

Circus Horses

Circus horses, the big dappled gray ones and the sleek, well-bred white ones and all the rest have successfully withstood the attack of the motor truck. It is reported that Ringling Brothers had an offer from a well-known automobile concern to furnish enough trucks to haul all the band wagons and various vehicles of the time-honored street parade, merely for the advertising which the concern would get from the use of its trucks. Ringlings declined the offer, so the story goes, and will continue to use the good, old-fashioned means of transportation that never fails, rain or shine, sand or mud.

Despite the high prices paid for good horses suitable for circus purposes it is not always an easy matter to pick up a large number of them. The ideal horse for the showman's purposes is a blocky, heavy-boned, short-coupled drafter weighing around 1,700 pounds. They must be well muscled and have good feet and legs. Where such a horse could once be purchased for $175 to $200 they now bring from $350 to $500 and for well-matched teams of dapple grays as high as $700 each has been paid.

TRUCKS INSTEAD OF HORSES? An interesting story published in *The Crow Bar* magazine in the November 1915 issue is reproduced here. The Ringlings toyed with the idea of replacing their Percherons with trucks in 1906. It was reported in the October 13th issue of *Billboard* that the big circus was dickering with a large truck manufacturer—a move that would cost them one-quarter of a million dollars. Neither of the above plans came to pass. The Percherons were finally put to pasture at the end of the 1938 season.

EVERYONE TURNS OUT. Campbell Bros. Circus parade comes down the unpaved main street of a midwestern town in 1908 as crowds watch.

QUICK STEPPERS. Coming back from parade, Cole Bros. team, pulling the United States bandwagon, found the incline presented a real pull. Note the fast stepping leaders as they hustle to keep the wagon moving. 1939.

40-HORSE HITCH. Barnum & Bailey Circus, in 1897, put this spectacle on the city streets wherever they played as part of their parade—a 40-horse team driven by one man. These were grade Percherons, bay with black manes and tails. It was an imposing sight and impressed on the city folk crowding the curbs that this circus was a truly big show.

LINING THE STREETS. In the big cities the crowds jam the curb to watch Sells-Floto Circus parade down the main thoroughfare.

PULLING THE BANDWAGON. A 10-horse team pulls the big, colorful, beautifully decorated bandwagon. The feather plumes on the horses' bridles were generally red in color.

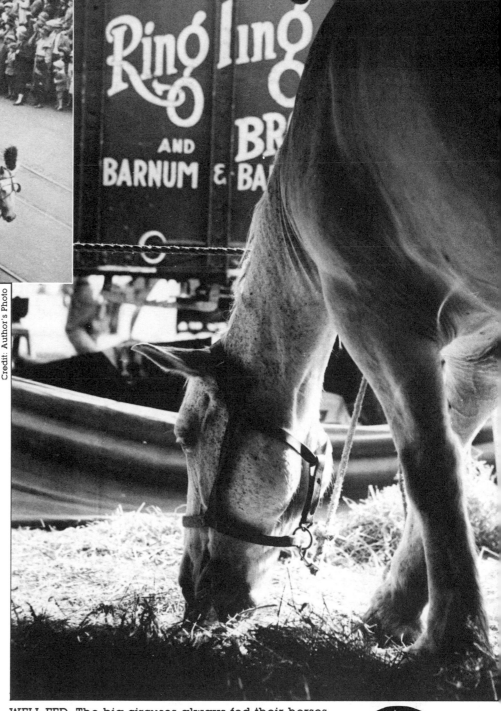

WELL-FED. The big circuses always fed their horses the best hay obtainable and grain three times a day.

EMPHASIS ON HORSES. In the era we are concerned with in this book, horses were a predominating feature of any circus performance. All shows advertised their horse acts forcefully as can be seen in the posters pasted on fences and barns and sheds a couple of weeks before the show arrived in town.

JUMPING THROUGH HOOP OF FIRE.   THE GREAT MILITARY DRILL FORMING INTO LINE.   MARCHING BY PLATO...

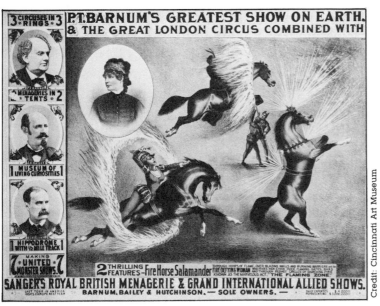

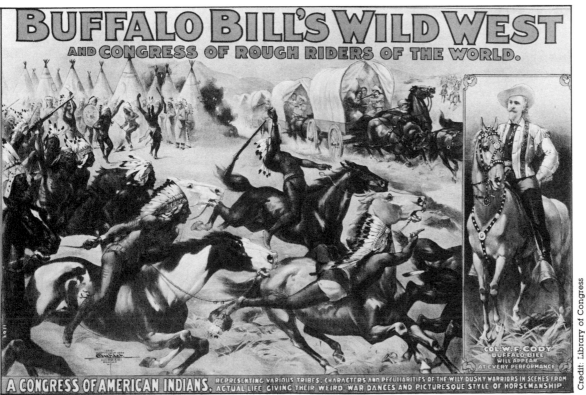

CHAPTER XVII

HORSES ON STAGE & AT FAIRS

EDUCATED HORSES were the rage in the horse age. The magnificently trained animals appeared at expositions, state fairs, county fairs, in theatres and on vaudeville bills.

These animals seemed to perform unbelievable routines as far as the audience was concerned. The people just never caught the cunning signals the trainers gave to these trained horses.

In 1882 Prof. George Bartholomew offered what he called "an Equine Paradox of 17 Educated Horses—all appearing at one time on the stage, the entire evening, entirely untrammeled by harness. They enter into the spirit of the entertainment with a zest, zeal, and gratification surpassed only by the pleasure of the audience—do everything but talk —understand over 300 commands comprising a vocabulary of over 1,400 words."

When Prof. Bartholomew played Washington, D.C., the Washington Post reviewed the act: "The National Theatre was crowded last night to standing room only and the frequent bursts of applause showed how thoroughly the audience enjoyed the novel entertainment styled an Equine Paradox—these horses appear to have almost human intelligence. They understand their own names and the names of their associates—they can do everything but talk and no undue force or whipping is resorted to. Prof. Bartholomew explains the methods he has adopted in their education and gives an entertainment with his dumb company far more pleasing than many rendered by speaking troupes. The Acts are unique and attractive."

A German trainer named Karl Krall had three or four educated horses that could tap out answers to arithmetic problems! One detractor tried to debunk his work in a published book by saying the horses only responded to subtle visual signals from the trainer. To defend himself, Krall then came up with a blind educated horse that stunned everyone.

Beautiful Jim Key, the celebrated educated horse, made the headlines in 1899, 1900 and beyond. He was considered the greatest of all Park attractions as he drew enormous crowds to Chicago's White City, Pittsburgh's Exposition, and Philadelphia's Woodside Park. The manager of Baltimore's Riverview Park said, "Jim Key is the greatest drawer and money maker we ever had."

This horse seemed positively ingenious as he picked out a number from a rack that correctly answered an addition or subtraction problem called in by someone in the audience. So subtle were his cues that the crowd actually believed Jim Key could pick out the red scarf worn by a child, who was one of several in a row but each wearing a different colored scarf.

Colonel Fred was billed as the horse with the human brain. In addition to solving arithmetic problems, Fred could spell, and could bang out a tune on a series of pendulum-like chimes.

The dark brown mare, Lady Wonder, was not only excellent at mathematics but was taught to spell. She would nudge blocks with letters on them and thus spell out the answer to questions asked. So beautifully trained was the mare, that scientists and journalists were baffled and never could detect how she got her cues.

Audiences, in big towns and small, thoroughly enjoyed watching these educated

VARIED ACT. An 1888 poster of Professor Bristol's educated horse act

PROF. D.M. BRISTOL'S EQUESCURRICULUM

PROF. D.M. BRISTOL

HORSE LAUGHING

THE HORSES GOING UP AND DOWN STAIRS

JOHNNIE SANBOURN
IN HIS
MARVELOUS SWINGING ACT

AT THE BLACK BOARD

POTOSKIE BLACKING THE PROFESSORS BOOTS.

THE COURIER LITH. CO. BUFFALO, N.Y.

HORSES ON STAGE & AT FAIRS

horses perform their astonishing routines.

There were other intriguing displays of horse flesh. The giant horses were big attractions. One such horse, Nebo, was advertised to be 20 hands at the withers (6' 8''). Nebraska Queen, a Shire mare, weighed 3,100 pounds. Big John, a sorrel Belgian, tipped the scales at 2,640 pounds and stood 19.2 hands. One of the biggest horses ever was Brooklyn Supreme, an enormous strawberry roan, advertised to be 3,200 pounds and 19-1/2 hands. This horse wore a 40'' collar when harnessed, and it took 30'' of iron to make one shoe.

All of these record-breaking horses were top attractions at county and state fairs.

Then there were many billed as the "smallest horse in the world''. Others attracted attention because of an extra long mane and tail such as Victor with his 18-foot tail and 21-foot mane. Barnum advertised a hairless horse.

The high-diving horses seen at the fairs were always truly spectacular. These animals were trained to walk up an inclined ramp then jump off into space, plunging into a portable tank of water. At the turn of the century, two Arabian horses, King and Queen, were presented by Prof. G.F. Holloway. Jumps of 30 feet were advertised and some promoters claimed that their horses dove from sixty feet in the air. Later, the Carvers were the big name in diving horses and had up to 10 horses in the troupe.

The general public enjoyed and admired horses and patronized all of these "off beat'' presentations.

PROGRAMME
— of the —

THE
Equine Paradox

Broadway and 35th Street.

Broadway and 35th Street.

Can you solve it?

EDUCATED HORSES.

Express Printing Company. Rochester, N.Y.

COMPLIMENTS OF THE EDUCATED HORSES.

To Our Patrons.

We desire to call your attention to the following points in our exhibit, points that will enable you the more intelligently to understand its true nature. Many of our patrons lose its best features by being carried away with the astonishing feats that the pupils perform. Wonderful as they are, it is not through them that we seek to merit your patronage and approbation; *but rather through the intelligent action of the pupils that grows out of them*. We claim an active intelligence for our school, an intelligence that grows with each succeeding day, proved and illustrated by the fact that no two exhibits are counterparts, in any sense, except in routine of work. Like in any school of chidren, each day brings out or developes new and peculiar characteristics, and it is due to this peculiar feature that a great majority of our patrons come many times to see the beautiful equines. It has been said, and with truth, that no one can see Professor Bartholomew's School without ever after not only loving a horse, but having a new respect for their kind. In other words, he has after years of patient, loving labor, elevated the horse to a higher plane, along which moves the progressive, the humane, the loving elements of life.

To describe the exhibit is beyond the power of any pen, no matter how facile; but to point out the various features, those worthy of the closest observation and study, is possible, and we therefore call your attention to them, in the order in which they come. Our school of horses understand over *three hundred* different commands, comprising a vocabulary of over *fourteen hundred* words. Any one of these commands they will obey as soon as given, stopping short in the execution of one to obey another given immediately afterwards. No matter how astonishing and incredible this may seem, it is none the less true, as we stand ready to prove in a special public exhibit whenever called upon to do so.

Programme.

THE FIRST BELL.
to School.—Introductory.
their way to School. In this opening act we find the pupils at
So skilled are they in this play

SECOND PART.

"Right this way, you Volunteers!"

The Drill of the Awkward Squad.—In this drill we find the pupils going through all of the intricate evolutions of veteran soldiers, and we challenge any of the crack regiments to go through as difficult maneuvers with as much precision as is found in the drilling of the Awkward Squad. Throughout the drill is found much to admire from the spectacular stand-point, the "obliques" being worthy of special mention.

The Battle.—"What! not a soldier left to defend our country's flag? My horses shall take their place!" A battle after the manner of men with horses as the contestants, is at least novel; but when once witnessed, we lose ourselves in wonderment. To them it is a genuine affair, and they enter into it with as much spirit and energy, as is possible to their natures. The flag scene is of special interest and beauty, making one feel like doubting that the transmigration of souls is a myth. It must be seen to be understood, let alone appreciated.

IN CONCLUSION:

The horses forming this School are, with three exceptions, blooded stock, costing outside of their keeping and education, over $20,000. Their education was undertaken without any idea on the part of Professor Bartholomew of making a public exhibit of their acquirements; but so astonishing were the results growing out of his discovery of the Pferter Garten system, that he was fain to alter his mind. To fit them for the stage took time, money and patience; but the looked-for result was at last attained, and to-day, through his ingenuity and unwearied perseverance, we have the pleasure of presenting to you, an exhibit as praiseworthy, as it is novel, interesting and amusing.

We solicit the patronage of the public from the humane, instructive, entertaining, and scientific standpoints; and we guarantee: 1st. There are no objectionable features to the exhibit; nothing, in fact, to offend the most refined lady. 2d. That it partakes of nothing old, but is amusing, entertaining and admirable from every standpoint.

Prof. GEO. BARTHOLOMEW, Prop.   F. M. COMSTOCK, Manager.

Cast of Characters.

THE PROFESSOR,	GEO. BARTHOLOMEW
A Knight Errant,	Chevalier
A Reasoner, from the Pacer Family,	M. Prince
The Head of the Church, but no jumper,	Pope IX
A Savage Chief, from Lower California,	Mustang
An Assine Judge of the Criminal Court,	Judge
A Whole Constellation,	Draco
The Hambletonian that takes a hand,	Brutus
How can you explain it?	Cæsar
Too Lazy for anything, and no jump,	Jim
As Awkward as Comical,	Bucephalu.
A Sorrel Nipper,	Gold-Dust
	Miss Nellie
	Miss Sprite
	Miss Abdallah
	Miss Beauty
	Miss Petite

SIXTEEN OF THEM

THE SECOND BELL.

...ging of the Second Bell we find the scene transformed from ...nfusion to one of perfect order and discipline, the scholars ...nselves in a becoming manner to properly receive their teacher. ...ould be closely watched by the auditor, for the great charm ...t lies in the little things, actions of expression, impossible to ...programme.

...Professor.
...phulas, take my hat and bring me a chair.
...e, the Monitor.
...and Brutus, the delinquents.
...ie, bring my Mail.
...e Graces of Abdallah.
...sic hath Charms.
...e Dancing of the Jig.—Bucephulas.
...quine Statuary.

...esar, in "How can you explain it?"—This act of Cæsar's ...plainable in any way unless you accord to the horse an active intel-...ce than any other, perhaps, on the programme, and we call your ...as it is evident that he not only understands right from left, but the ...nce between a circle and a figure eight. To make this act of more ...ordinary interest, the audience should express their desires promptly ...istinctly.

The School at Recess.—More intelligent action grow out of ...scene than any other, perhaps, on the programme, and we call your ...ntion to the following facts relative to it: If any pupil fail to promptly ...ond to the command of the Professor, you will observe those in his ...r vicinity force him out by biting, kicking, or any other means in their ...wer. Such of them as do not like to jump will beg off like any child ...ected to perform a disagreeable task. All of the wonderful feats per-...rmed in this scene are executed by the simple word of command.

Beauty and the Barrel.—"Does it hurt your nose?"
Nellie and the Barrel.—"Over the plank. This graceful and very ...difficult feat is a fitting finish to Beauty's comical act. It is a beautiful ...picture, well worthy of the applause always given.

Reasoning from Cause to Effect.—Prince and Pope. In this ...feat the most stubborn doubter is convinced that horses do possess the ...power of reasoning from cause to effect, though the fact is often lost sight ...of because of the sudden transition from the scientific to the ludicrous and ...absurd.

See-Saw.—Prince and Pope.
Grand Tableau.—In this Tableau we find the whole School, ...at the word of command, taking the various positions assigned to them ...and holding them, no matter how difficult they may be, until ordered to ...their places. Perhaps as much intelligent expression and action will be found ...in this tableau as could be found in its like if formed by human beings.

The Murder. Circumstantial Evidence. Guilty or not
Guilty. The Trial. The Jury Retire. The Verdict. "Horse
...can never be questioned again.

EDUCATED HORSES. The four-page program distributed by Professor George Bartholomew to his patrons in 1888.

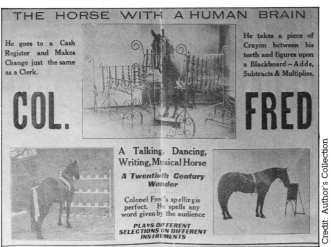

WONDER HORSE. Col. Fred performed in the 1920's. He was a "Twentieth Century Wonder that could talk, dance, write and play music."

MINIATURE. Jerome Selleck presented his miniature horse at Sans Souci Park in Chicago in 1906. It was billed as the "Smallest Horse In The World".

EDUCATED HORSE. Jim Key was owned by A. R. Rogers and performed at White City Park, Chicago, in 1906, as well as other locations.

FLOWING LOCKS. A horse with a mane like this was a freak of nature, but they were a great attraction on the County Fair circuit. Circus side shows also displayed such animals. Ringling Bros. had a long-maned horse in 1894. Buffalo Bill's Wild West exhibited a horse with a mane growing out of the middle of its back, and Barnum & Bailey Circus presented a "hairless horse" in the 1890's.

VICTOR. This ad appeared in the New York Clipper, September 16, 1899.

STUNT HORSE. High diving horses were very popular at the County and State Fairs. Here Eunice Winkless rides her horse into a tank of water at Pueblo, Colorado, July 4, 1905.

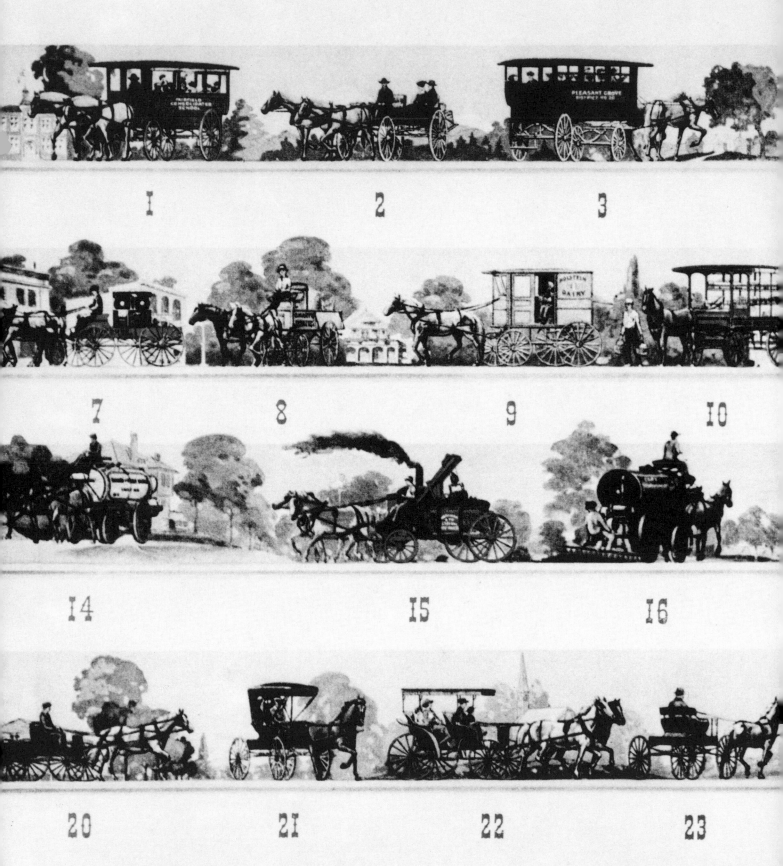

From ice wagons to elegant surreys, the Studebaker lines presented virtually the whole panorama of horse-drawn equipage used by the populace in the decades just before and just after the turn into the 20th century. Farmers, urbanites, commercial users and municipal officials chose from thick catalogues—and if they did not find exactly what they wanted, they could have it by special order. Said the company: "We are prepared at all times to submit designs covering any class of vehicle known to the carriage and wagon trade." Studebaker vehicles were noted for their style and smart color, as well as traditional durability. Those above were shown in a brochure of the time to indicate the breadth of the company's experience.

HE HORSEDRAWN VEHICLE

Studebaker — ESTABLISHED 1852 — BUILDERS OF VEHICLES FOR EVERY PURPOSE

SOUTH BEND, IND.

1. Curtain side school bus
2. Two-seat park wagon
3. Glass side school bus
4. Two-horse lumber truck
5. Four-horse log wagon
6. Six-horse log wagon
7. Half platform delivery wagon

8. Platform spring furniture wagon
9. Milk wagon
10. Roll curtain top delivery wagon
11. Standard two-horse platform truck
12. Ice wagon
13. Coal wagon
14. Sewer flusher

15. Contractor's wagon
16. Road oiler
17. Street sweeper
18. Sprinkling wagon
19. Garbage wagon
20. Village market wagon
21. Izzer buggy

22. Canopy top surrey
23. Business wagon
24. Panel top grocery delivery wagon
25. Farm wagon
26. Auto seat Studebaker buggy
27. Three-seat platform wagon